Wylam
200 Years **of Railway History**

Wylam
200 Years of Railway History

George Smith

AMBERLEY

First published 2012

Amberley Publishing
The Hill, Stroud
Gloucestershire, GL5 4EP

www.amberley-books.com

Copyright © George Smith 2012

The right of George Smith to be identified as the Author
of this work has been asserted in accordance with the
Copyrights, Designs and Patents Act 1988.

All rights reserved. No part of this book may be reprinted
or reproduced or utilised in any form or by any electronic,
mechanical or other means, now known or hereafter invented,
including photocopying and recording, or in any information
storage or retrieval system, without the permission in writing
from the Publishers.

British Library Cataloguing in Publication Data.
A catalogue record for this book is available from the British Library.

ISBN 978 1 4456 1077 1

Typeset in 10pt on 12pt Sabon.
Typesetting and Origination by Amberley Publishing.
Printed in the UK.

Contents

	Introduction	7
One	Coal in the North East	9
Two	The Growth of Wagon-ways	20
Three	Wylam Colliery & Christopher Blackett	29
Four	William Hedley, Timothy Hackworth, Nicholas Wood & George Stephenson	38
Five	The First Wylam Experiments	45
Six	The Development of the Steam Locomotive	51
Seven	*Puffing Billy*	58
Eight	Stephenson's First Locomotives	65
Nine	*Wylam Dilly & Lady Mary*	71
Ten	Stephenson, Tennant and the Stockton & Darlington Railway	78
Eleven	The Wylam Public Railways	85
Twelve	Timothy Hackworth and the Locomotive Industry	95
Thirteen	The Wylam Boom Years	104
	Epilogue: The Wylam Legacy	111
	Acknowledgements	122
	References	123
	Principal Bibliography	126

Introduction

It is commonly believed that teenagers are spotty and inconsiderate and think of nothing other than sex and booze. When I was a teenager I wasn't like that. At least some of the time I thought about trains. Admittedly the rest of the time I thought about nothing other than sex and booze, but I was neither spotty nor inconsiderate. It was everyone else.

On the days when trains dominated, and I had sufficient disposable income to make the trip, I caught the train to Darlington Bank Top station. Darlington is of course on the main line between London and Edinburgh and by going there I saw more of the sort of locos I thought worth seeing. My native town of Hartlepool specialised in rust buckets which limped and wheezed out of West Hartlepool shed and whose value was unappreciated until they were all gone – alas-alas. To get to Darlington I caught two trains, changing on the way at the windswept outpost of Thornaby, and caught the Middlesbrough crawler which terminated at buffer stops, under a stone plinth, on which stood two ancient locomotives. One was the *Derwent* about which I recall nothing, apart from the name, and the other was *Locomotion No. 1* which thankfully means more to me now than it did then. I shall forgo the opportunity to tell you about the *Derwent* [1], as it isn't pertinent to the story I am going to relate, and will concentrate instead on *Locomotion*. At the time in the mid-sixties I am talking about both steam locomotives and railways in general were in terminal decline. The great non-stop steam hauled express trains from Scotland, such as the *Flying Scotsman* and *Elizabethan*, were still around but in their death throws. We, both the spotty and the immaculately groomed (such as myself), stood on the south end of the station platform in order to catch a last glimpse of them as they hurtled into oblivion in a final flurry of noise and steam. It always rained then so on those rare occasions when there was a lull in the traffic I sneaked away to dry out over a cup of unidentifiable soup in the station snack bar. To do this I again had to circumnavigate *Locomotion*. Over the course of five years I must have passed it fifty times. So what did I think about it, if anything? Well not much to be honest. To me it was a relic of a bygone age, a statue, an artefact – no different to the stone images of self-satisfied and long forgotten dignitaries gazing blandly down from pigeon spattered columns on every high street. I could no more relate Locomotion to the screaming monsters I came to see than I could the tea urn in the snack bar; at least the tea urn seemed a living object spouting steam – well sometimes anyway.

Another twenty years would pass before I met *Puffing Billy*. I was living in Windsor and had dragged my reluctant children for a day out at the Science Museum in South Kensington. There amongst the high tech detritus was a strange wooden barrel on wheels. On closer inspection it turned out to be a steam locomotive. I had never thought of *Locomotion* as being particularly small; elevated as it was on a plinth and separated from other objects that might have given it scale, but *Puffing Billy* looked like a Victorian sewing machine someone had abandoned in a modern furniture shop. Before getting dragged away to look at space rockets I had time to glance at the information board. It turned out that *Puffing Billy* was the oldest surviving steam locomotive in the world being built nearly two hundreds year before my child friendly excursion to South Kensington. I had just enough time to read that the engine originated in the north east and was designed by someone I had never heard of before I was hurried away. So *Puffing Billy* left no more lasting impression than *Locomotion* had once on my teenage sensibilities. I know better now.

Locomotion and *Puffing Billy* are important. They are fundamental to the story that follows. The men who designed and built them, George Stephenson and William Hedley, hailed (if that's what people actually do) from a tiny village in Northumberland called Wylam. This too doesn't mean much in itself – after all we all have to come from somewhere – were it not for the fact it turns out they were not the only great railwayman to 'hail' from there. Their Wylam contemporaries included Timothy Hackworth, perhaps the world's first great locomotive builder and Nicholas Wood, the first person to recognise the international importance of railways and document their origin and history whilst simultaneously providing a template for their reliable construction. So what was it that made Wylam special? Indeed, what combination of circumstances and events in this unassuming colliery village transformed pit men into pioneers? I hope that what I am about to relate goes some way to providing the answer.

Before jumping head first into the story however I would like to thank a number of people who provided invaluable help along the way:- Denis Peel and Philip Brooks of the Wylam History Society for their help and courtesy; Alison Kay of the National Railway Museum for permitting an early look at the Hackworth Family Archive; Dr Jill Murdoch of the Institute of Railway Studies at the University of York for her assistance and encouragement; the staff of the British Library at St Pancras and the National Archive at Kew for patiently helping someone hopeless with technology; the staff of the Northumberland County Archive at Woodhorn for the specific information relating to the Blackett family and my good mate Tom Horsburgh who knows more about railways than I ever could and who accompanied me during much of the research work and liquid refreshment interludes so essential to the amateur historian. Finally I would like to thank my family, Niall, Owain and Holly and in particular my long suffering wife Maggie who had to put up with a lifetime's accumulation of railway junk and provided the necessary criticism of my work that kept it focussed and relevant.

CHAPTER ONE

Coal in the North East

There is an engraving of the village of Wylam as it appeared in 1836. A train comprising a ragbag of wagons and passenger coaches approaches the village along the southern bank of the Tyne; its double decked coaches heaving with passengers and the open upper decks a sea of heads. On the northern bank is Wylam. It consists of slag heaps, the occasional cottage and a forest of smoking chimneys. Wylam was a thriving industrial town then, complete with a colliery, an iron works and an engineering foundry. Fast forward to today and all the industry has gone – along with the chimneys, the smoke and the double deck passenger coaches. Wylam has become a leafy backwater; home to commuters employed in the urban sprawl of Newcastle and Gateshead. Where did all this industry come from and where did it go? We know that when the engraving was made the river ran black with colliery waste and the hills echoed to the clang of hammer on metal yet the Tyne there is now crystal clear. Wylam has become the haunt of anglers seeking a weekend break in the country. After more than 200 years the wheel has turned full circle. Wylam is now the quiet backwater it originally was. So what happened?

The railway in the engraving is the former Newcastle and Carlisle Railway (N&CR); a line built to connect the east and west coasts of Northern England and amazingly – given the wholesale destruction of the railway system by Dr Beeching – it is still there. Hourly trains from Newcastle still stop at Wylam's tiny Victorian station. The former N&CR line, following the beautiful Tyne valley up and over the Pennines to the Irish Sea, remains a vital part of our national rail network. However, it wasn't the only railway in the village. Somewhere among the chimneys was another, perhaps historically more significant railway, which ran from the centre of the village to meet the Tyne five miles east at Lemington. This book is about that railway and the role the tiny village of Wylam played in the birth of railways. Wylam has its own unique place in railway history and it owes it all to the presence of coal. It was coal that would be the driver for all that followed.

People had been mining coal in north east England for two thousand years. There are references to the use of coal for fuel in correspondence from the

Wylam Village in 1836.

Roman Army stationed on Hadrian's Wall two miles to the north. The Romans were even known to have operated their own coal mine, presumably manned by local slaves, at Benwell in Northumberland. The geology of the North East is such that coal, in one and two metre layers, outcrops on the surface of both land and seabed. I recall from my own childhood in Hartlepool a daily parade of men like my father, pushing bicycles with sacks strung over the handlebars, heading to the beach to gather the harvest of 'sea coal' that accumulated in thick black drifts at low tide. North eastern coal was always historically called 'sea coal' regardless of its origins so it is likely that the coal, washed up on the sands each day, constituted the first commercial export from the region, since its sale merely required collection and distribution. Beach deposited sea coal isn't perfect – as anyone who has ever used it knows – the quality is variable at best; usually it contains stones, shells and other beach detritus. This could explain the widespread reluctance towards its domestic use in the middle ages. However this didn't just apply to sea coal from the north east. In the Middle Ages there was a preference, even in coal producing parts of the country such as the Midlands and Kent, for charcoal because low grade coal contains impurities which, when burnt, produce the poisonous yellow smog that so damaged people's lungs in the 1950s.

The preference for wood or charcoal, at least for fires in the home, persisted until the end of the sixteenth century when an imminent threat of invasion from the continent made the Tudor monarchs undertake a major programme of ship building. As the forests fell, to be transformed into the hulls and masts of ships, so the source material for charcoal became scarce and its users were obliged to look elsewhere for a suitable, locally available alternative. The solution lay in abundance right beneath their feet. It transpired that the UK is singularly blessed with coal with vast deposits spread throughout the country.

There were, needless to say, problems to overcome before coal could be used as a straight replacement for charcoal. Unfortunately not all coal is the same. Certain types contain high concentrations of undesirable elements such as sulphur and phosphorus, the elements indeed that caused the London smogs. Other forms are so dense they are difficult to ignite, and once alight become hard to extinguish. Some coal also decays far too rapidly; evolving large volumes of flammable methane gas making it dangerous to work other than in the open air at the surface. Then there are the coals that are very friable, rendering them difficult to handle and aesthetically undesirable. This type the miners called 'small coal' and for a long time was a waste that accumulated in vast heaps around the pit head. The worst coal of all however, at least from the point of view of the land owners, was the sort that was too expensive to mine either because it either lay in seams too deep underground or because it was so far away from potential customers that the transport costs outweighed any potential profit from its extraction.

Taking into consideration the above criteria, north-east coal turned out to be the best; the 'Geordies' were doubly lucky because right on their doorstep there was an awful lot of it. The north-east coal field is 50 miles long and 25 miles wide extending across most of Northumberland and Durham and many miles out under the North Sea. The proximity of the sea was also a blessing to mine owners. It meant they had a cheap way of getting their coal to the rest of the world – even if this was restricted to using boats moored on the short deep navigable stretches of the estuaries of the three principal rivers, the Tyne, the Wear and the Tees.

In consequence North-East England became synonymous with coal from the days when serfs gathered it at the command of knights on horseback. In 1239 Henry III granted a charter to 'the good men of Newcastle ... to dig coals in the common soil of the town, without the walls thereof'.

Commercial exploitation of north east coal however didn't start in earnest until a hundred years later. The original source of the black gold was the monasteries of Northumberland who allowed the establishment of mines on their land – for a suitable fee obviously – the coal being mainly sold for use in iron foundries where air pollution was less a consideration. Over the following years commercial mines opened up right across Tyneside, with the coal exported from Newcastle, then the western navigable limit of the River Tyne. Demand increased significantly in the seventeenth century when the reigning monarch traded in his elderly and barren wife for a sportier model, a move of which the Catholic Church seemed for some reason to disapprove. To overcome the embarrassing theological difficulties, Henry VIII, as we know, declared himself head of the English church and was thus able to acquire all church land, including that which had once belonged to the monasteries. This he distributed among his political cronies; following which there was a sudden spurt in coal production, as the new occupiers of church sought to maximise their ill-gotten gains. In 1622 alone nearly 15,000 tons of coal was exported from Tyneside and it must have been annoying that by this time most 'open cast' coal seams were already worked out. Consequently, underground mining, in the sense we understand it today, truly began; the roofs of the tunnels that burrowed into the hills being propped up with timber or left supported on undisturbed pillars of coal.

There were still huge profits to be made in the years following the end of the Tudor dynasty but this was a turbulent time for the North East. During the Jacobite wars there were frequent cross border raids and local militias were established to defend the expanding network of collieries. Mining became the industrial cash crop that financed the bloody battles between the English and the Scots. A measure of the extent of this conflict is that one (at least) of the worked out mines ('Stella Grains Lease' in Northumberland) was used, according to local sources, as the dumping ground for 'numbers of bodies from the Border Wars'.

Whatever else was going on at the time there was good money to be made from digging coal.

Deep mined coal, like open cast, was the property of the lucky land owner on whose land the coal deposits lay. Land owners rights ended, nevertheless, when the seam extended beyond the landowner's boundary regardless of how deep the seam was below the surface. In addition, whenever transporting coal meant crossing a neighbour's land, the neighbour could charge for access across his land. This rent was called 'wayleave' and was initially a purely northern phenomenon. It certainly came as a shock to the brother of the Lord Keeper of Guildford, Roger North, who ran into it for the first time during a trip to Newcastle in 1676. The main custodian of church land was now the Church of England and the Church, particularly in County Durham, owned vast tracts of land, most of it moorland with little agricultural value. If they couldn't farm the land then they would find some other way of profiting from it. The Clergy proved particularly greedy when it came to wayleave. Tomlinson, in his book on the North Eastern Railway, says that the Bishop of Durham made more profit from wayleave rents in 1716 from one colliery than did the actual owners of the mine. The same clergy, many years later were instrumental in the bankruptcy of the Stanhope and Tyne Railway in County Durham, and along with it their most famous shareholder, Robert Stephenson, for exactly the same reason.

Working conditions for miners were unvaryingly bad. They were only paid for the coal they actually produced. [2] Consequently, any time spent on health and safety issues, such as installation and maintenance of roof supports, was so much wasted effort. The situation was made even more wretched by the routine exploitation of the miner's wives and children who were press ganged into duties ancillary to the heavy labour at the coal face. To get the coal to the wagons at the surface the miners laid wooden planks end to end to form a simple road. These roads were called 'barrow-ways' and it was often the miner's wives and children who dragged the 'toads' or sledges of coal along the barrow-ways from tip face to pithead, which were then raised to the surface using winding 'gins' or pulleys. Women usually acted as the haulier. Naked to the waist, they were fitted with a harness over their shoulders and around their stomach with a chain that ran between their legs fastened to the loaded 'toad'. Children often assisted, pushing the sledge along from the rear. This practice continued right through to the middle of Victoria's reign and was extensively commented upon by Engels in his seminal book about working class conditions, published in Germany in 1845. On more productive mines the gins were worked by horses or water wheel if there was a nearby stream, but on marginal pits the gins were worked by youngsters or old men. Miners' children started working down the pit as soon as they could walk, the youngest being given the job of opening and shutting the ventilation doors and partitions; sitting alone in the dark for hours at a time.

The key men working the face were known in the north east as 'hewers'. [3] They needed to be athletes to endure the terrible working conditions. The physically demanding work, under cramped conditions, eventually produced, by natural selection, the typical short stocky northern 'pitman' George Orwell describes in his book *The Road to Wigan Pier*. Orwell gives a lurid account of just one day he spent with men working underground in a northern coal mine. In the mine he visited the coal face, which was more than a mile from the pit head. Since the miners weren't being paid for non-productive time they started the shift by sprinting head down from pithead to coal face loaded up with their shovels, lamps and other paraphernalia. The combination of a wet slippery floor and a low roof meant that the tall gangly Orwell kept slipping and banging his head.

It was over an hour after starting the shift before the miners actually started cutting coal – an hour for which they received no payment. Although in his early thirties and physically fit, by the time he reached the tip face, where the real work began, Orwell was exhausted. He said of his short time down the mine,

> I am not a manual labourer and please God I shall never be one, but there are some kinds of manual work that I could do if I had to. At a pinch I could be a tolerable road sweeper or an inefficient gardener or even a tenth rate farm hand. But by no conceivable amount of effort or training could I become a coal miner; the work would kill me in a few weeks.

This was in 1936; a hundred years earlier conditions were much worse.

In consequence, life expectancy was short; hard labour being only one cause of early demise. Coal, in the absence of air, decays to produce the flammable gas methane. This wasn't a problem for open cast miners since concentrations of methane never accumulated to the point where they became explosive (5 per cent to 10 per cent by volume in air), but in the confined tunnels and galleries of underground pits, high concentrations of gas were an ever present fire hazard. Methane, or 'fire damp' as it was called, only needs a spark to set it off. This, unfortunately, was something miners carried with them at all times since they needed a light source just to see what they were doing. The open flame of the candle they wore in their hat was the cause of many a pitman's death until the arrival of the safety lamp – more of this later. With this ever present menace it is no surprise to learn that explosives were rarely used in early coal mines. However, if this reduced the risk of death by explosion, it did little for the toll on the 'hewers' bodies which were subject to the additional labour needed to loosen the coal from the tip face. Even when methane wasn't a danger, carbon dioxide often was. Carbon dioxide despite being naturally present in exhaled air, rapidly induces narcosis and subsequently death when present in high concentrations. Tasteless, odourless and heavier than air, it lay in invisible lethal fogs in the deepest recesses of the mine. It was the potential presence of carbon dioxide or 'miners damp' that led to the use of caged birds to act as a warning; the birds reacting to non-lethal (to humans at least) concentrations of the gas.

If by some miracle miners survived their time underground their lives were often cut short by illnesses such as 'black' lung diseases caused by continual ingestion of coal dust. The working life of the average coal miner normally ended in his late thirties. The miner was usually dead before his fiftieth birthday. It is not for nothing that Friedrich Engels, who knew a thing or two about the short miserable lives of working people, said about coal mining: 'In the whole British realm there is no occupation in which a man may meet his end in so many diverse ways as in this one.'

By the time the mining industry had perfected its own unique expertise in killing or maiming its employees it had become a specialist industry with its own peculiar terminology.

Day to day management at coal mines was handled by 'Viewers'; the term reflecting the 'overseeing' nature of the job. The level of pay 'Viewers' received, although more than the labourers, was variable, along with the differing nuances and complexities to the work they were expected to undertake at individual pits. In addition to man management 'Viewers' also had technical responsibilities and often acted as the engineer for one or more groups of mines. Viewers controlled the wages paid to the pitmen and hence had the power of life and death over them. They would often reduce the amount the men were paid on the flimsiest of excuses, such as the presence of too much coal dust amongst the coal being loaded into wagons or turning up a minute or two late for work. In consequence they weren't liked much. Each mine also had a resident 'blacksmith' whose duties, whilst largely self-evident, would soon include the mechanical know how needed to maintain and operate steam powered water pumps.

Once the coal reached the surface it was segregated by hand with the more saleable bigger chunks transported to the customer via sea going collier boats waiting on the tidal reaches of the river. When the pits were located some distance from these boats this was a time consuming business since, in the early years of commercial mining, coal was only moved in sacks loaded on the backs of pack horses. So, by the mid-eighteenth century, there emerged the keel boat (or its southern equivalent the 'lighter'); typically, a flat bottomed sailing boat 12 metres long and 6 metres wide. The sails were a mixed blessing. On the Tyne, for example, they had to be taken down whenever the boat went under low bridges. The keel was manned by two or three men and handled around 20 tons of coal. Because of its shallow draught, the keel could venture much further up river than the sea-going collier ships. It thus acted as an intermediary transport facility, reducing the distance between coal field and collier vessel that required the use of pack horses. By the beginning of the eighteenth century there were an estimated 400 keels on the Tyne alone. The loading of keels took place at coal staiths. These, in their simplest form, were wooden docking platforms on the river bank provided with chutes into which coal could be tipped and

Tyneside coal staith *c.* 1800.

then directed into the holds of the keels. Later refinements included piers that jutted out into the river (or sea) at some point on the river where a collier ship could be fully loaded without risk of running aground at low tide. Following the advent of railways coal wagons were soon modified to enable the floor to fully open and jettison the load. The staiths were then referred to as 'coal drops' where coal could be delivered into the holds of ships without further manhandling.

The geology of coal deposits on Tyneside was such that seams of coal were often sandwiched between layers of water permeable rock such as limestone. Consequently, when mines penetrated far beneath the surface, flooding became yet another hazard for miners to overcome. Since a flooded pit has no value this was an issue considered more seriously than health and safety by mine owners. The solution to this problem provided the first major stepping stone on the road to steam locomotion.

The 1700s was the century of the scientist. In Swift's *Gulliver's Travels*, written in 1726, Lemuel Gulliver visits a flying island called Laputa, inhabited only by technophiles. Swift's tale was a thinly disguised lampoon of science and scientists. His target, in particular, was the members of the Royal Society who included in their ranks such notables as Isaac Newton and Robert Hook. Swift's parody struck a chord with his readers because it was widely suggested at the time that science was going to solve all the outstanding mysteries of the world; as that was what the leading scientists of the day themselves believed. Britain was at the centre of an explosion in scientific knowledge, the underlying causes being social and political; social, in terms of the reduced influence of the Church compared to our European neighbours and political because of the stifling impact on science on the opposite side of the pond from such disastrous upheavals as the French Revolution. Consequently it was the British that made the major advances in the forthcoming industrial revolution.

The technological breakthrough driving all this was the arrival of the steam engine.

There are many claimants for the engine's inventor but, like most inventions, it is neither possible nor helpful to single out one particular individual. For each original idea there are a thousand incremental improvements that convert the concept to practical reality. The principle that machines could be powered by steam was not new in itself. It was self-evident that a tiny amount of water produced a lot of steam (in fact a table spoon of liquid water generates a dustbin full). That steam could also be harnessed to perform useful work was also well understood, even by the ancient Greeks. The philosopher Hero made a simple engine using steam ejected from a tube to rotate a drum on a spindle. However there were no practical applications until scientists began searching for an effective way of dewatering flooded mines.

Mechanical methods for lifting water had been around from the days of the Pharaohs but all relied on simple baling techniques in one form or another. These were slow and inefficient and incapable of dealing with the sort of rapid flooding that deep mines were subject to. What was needed was a device that operated continuously with minimal supervision. Hand operated pumps had been used in the past to deal with small intrusions of water but a pump was about to appear that required no human (or animal) assistance to make it work.

Pioneering experiments were conducted by Thomas Savery towards the end of the seventeenth century. Savery was able to show that a vacuum could be produced in an airtight container if you displaced all the air in the container with steam then rapidly condensed out the steam with a jet of cold water. In Savery's machine, patented in 1696, the vacuum created was used to pump flood water from mines via a pipe connected between the flooded area and the vacuum tank. After the water in the tank had been discharged it was then resealed and the process repeated. A later engineer, Thomas Newcomen, worked out that if you replaced Savery's vacuum tank with a cylinder fitted with a piston then the piston would be forced into the evacuated cylinder by atmospheric pressure alone. If steam was then reintroduced into the cylinder the vacuum would be broken and the piston would return to its starting position. He used this principle to make a water pump by connecting the piston to a baling bucket via a pivoting beam, in the manner of 'nodding donkeys' you see on oil wells today. In this way a bucket of water could be lifted and emptied on each piston stroke.

Newcomen's water pumping engines proved very successful and were soon installed in mines throughout the country. Unfortunately they suffered from one drawback; they used a lot of fuel. This was because the vacuum cylinder had to be alternately heated and cooled which meant useful heat was being wasted on every piston stroke. This wasn't a problem at coal mines, given the abundance of waste ('small') coal to hand, but was a big issue in, for example, tin and copper mines that had to buy in any fuel that was used. It was left to the face on the fifty pound note, James Watt, to square the circle. In Watt's pump the steam in the cylinder was made to discharge to a completely separate chamber before being made to condense. It was therefore possible to keep the primary chamber, in which the steam had been first injected, insulated and maintained at high temperature. Watt increased efficiency further by utilising both sides of the piston for steam injection and condensation, thereby doubling the effectiveness of the pumping engine.

Despite these advantages, Watt's early experiments were only moderately successful because the engineering expertise and materials needed to build his engines wasn't available at the foundry where he worked in Scotland. It was therefore fortunate that he met Matthew Boulton, who had both the financial resource and technically competent workforce at his factory at Birmingham, to make Watt's dreams reality. With Boulton at his side Watt took the reciprocating piston principle an important stage further. He converted the simple up and down piston movement into circular motion by connecting the piston to a wheel. This could then provide power to operate all the pulleys and lathes already in use in factories including those owned by Boulton himself. The next logical step was to use the piston to turn the wheels of a moving vehicle and it wasn't long before

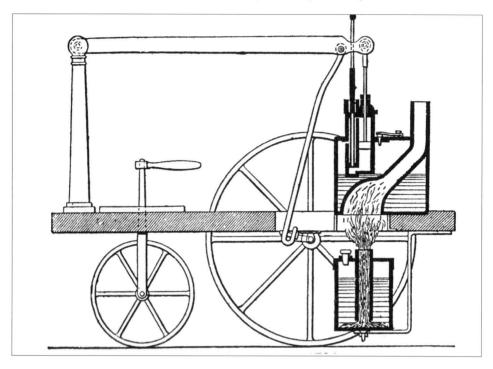

Above: William Murdoch's high pressure steam model engine.
Left: Richard Trevithick.

Watt started tinkering with such machines. It would be a rival scientist, Nicholas Cugnot in France, however, who first demonstrated the principle. Unfortunately scientists weren't the flavour of the day after the French Revolution and Cugnot kept his head down and took his work no further. Watt too was having problems. Heavy metal boilers on wheels required more power to move than Watt's low pressure steam engine could supply. The obvious answer was to increase steam pressure but Watt's first high pressure boiler trials scared the life out of him. He summarised his concerns: 'The danger of bursting the boiler, and the difficulty of making the joints tight...'

After one or two worrying scares he became convinced that the risks associated with high pressure boilers just weren't worth the candle. He even went so far as to curtail the research of William Murdoch [4], Boulton's foreman at the Birmingham factory, who had already constructed a prototype locomotive using high pressure steam. Murdoch, however, had by this time successfully tested out his model on the roads of Redruth where he was sent to supervise the installation and maintenance of Boulton and Watt pumping engines. His experimental machine, spitting fire and steam, was reputedly seen one night by a local priest who ran away believing it to be the devil incarnate; the first of a number of similar apocryphal stories associated with early steam locomotives. Following Watt's embargo no further work on high pressure engines was carried out by Murdoch, and Watt confined himself for the rest of his days to making minor improvements to his engines primarily to protect his patent. In the centre of Birmingham today, opposite the Symphony Hall in Broad Street, there is a statue painted in gold. The statue shows three men in animated discussion. The men were considered by the City fathers to have contributed more than anyone, before or since, to the development and emergence of Birmingham as the UK's 'Second City'. The men so honoured are William Murdoch, Matthew Boulton and James Watt.

The development of the steam engine might have ended there were it not that during Murdoch's fifteen years in Cornwall he met and befriended a young engineer already building a reputation for improving the efficiency of pumping engines. This young man had started out working on Watt's machinery installed in his father's tin mines but, while undergoing the engineering training provided by Murdoch, the older man's enthusiasm for high pressure steam rubbed off on his apprentice. The young man was Richard Trevithick and it was Trevithick who would go on to carry Murdoch's ideas onward and upward; along the way becoming the 'father' of the steam locomotive.

CHAPTER TWO

The Growth of Wagon-ways

At the eastern end of the oldest railway bridge in the world there is a cobblestone plinth and standing on the plinth is a wooden truck on a short length of wooden rail. The truck has four flanged wheels and is fitted with a crude wooden brake; looking like an outsize version of a child's toy – the sort you see in Early Learning Centres. Although made out of wood, the wagon has all the attributes of modern trucks yet is of a type that worked local wagon-ways three centuries ago. If immediately recognisable to modern eyes, it was not necessarily typical of its day. In those far distant pioneering years there were countless variations on this basic theme including some wagons that were little more than wooden hand carts. Nevertheless they all had one common feature. They all operated on rails.

So where did the wagon ways come from?

To find out we have to go further back; 1,300 years further back to be exact, to a time when the Roman Army was getting ready to abandon the garrisons on Hadrian's Wall. When the legions marched away on roads they had built themselves, the ability to make decent serviceable roads marched with them. Their successors, the Saxons and Normans, had little interest in wasting valuable raping and pillaging time on such pursuits and the basic packhorse and drove roads they created were little more than muddy tracks hacked out of the prevailing forests. Unlike the Roman highways these highways sought the line of least resistance, using the natural contours of the land. When faced with an obstacle they invariably went round it. Unlike the Roman roads, there was little in the way of surface dressing. The only cover the roads received was logs, which they laid down over particularly marshy sections. Consequently they were of little practical use for carrying any vehicle heavier than a loaded horse and cart. And so, when sixteenth-century colliery owners in North East England needed to shift coal in bulk they looked for a harder wearing and level surface on which their wagons could run.

The obvious answer was to follow the Roman model, remove any intervening obstacles, and lay down a decent hard wearing stone surface. It would have been easy to do. They had plenty of tough granite available nearby yet, surprisingly,

Coal wagon from Tanfield wagon-way.

they opted for simple railroads built with wood. These wooden wagon-ways had, in fairness, already been used in Germany from the sixteenth century onwards and German engineers, mining copper at Keswick, were already using them by the beginning of the 1700s. The wagon-ways the German engineers built consisted of wooden rails nailed to wooden sleepers. The four-wheeled wagons used on them were fitted with conventional cart wheels and, to prevent them slipping off the track, they attached wheel retaining strips to the inner edge of the rails.

The wagon-ways were an immediate success and the first one designed specifically for coal trucks was laid down at Beaumont pit at Blyth the following year. If an improvement on what went before the new roads brought with them their own set of problems. Inevitably with the heavy loads carried, the wooden rails soon fell apart. Since whole sections of track then had to be expensively replaced, the first modification was to lay a second wooden rail over the first. This could then be easily replaced when it wore out without having to take up the whole road. There was, however, still another problem to overcome. If the wheels of the cart were kept on the rail by a vertical edging strip then mud, stones, ice etc. could accumulate in the 'L' shaped recess between the edging and the rail surface. This led to regular derailments. What was needed was a rail with a round top that could then continually shed debris. The result was rails with a

Early Newcastle wagon-way.

fish shaped cross section and no edge strip. These, however, required a different sort of wagon wheel; one with its own vertical retaining strip to prevent lateral movement. This resulted in the flanged wheel adopted by the Tanfield Wagon-way and subsequently used on virtually every railway today.

The final change, which took fifty years to fully implement, was the replacement of wooden rails with their iron equivalent. This meant either swapping the wooden rail directly for its iron counterpart – an expensive option given the cost of iron at that time – or retaining the 'L' shaped template and plate over the wood with a thin veneer of metal. Both types of rail had their advocates. The iron rail devotees pointed to a longer working life for their rail and a smoother surface for the wagon wheels to move on. The 'L' shaped rail fans argued that their 'plate' rails cost less, since they used less iron in their construction, and that the wagons required no expensive modifications. They also had the advantage that the same wagons could be taken off the rails, if necessary, and used without alteration on the highway. For many years both types of rail existed side by side.

This was the situation at the time the mine owning co-operative known as the 'Grand Allies' bought rights to work coal on a site six miles to the south of the Tyne, north of the village of Tanfield, County Durham. The 'Allies' were a group of wealthy landowners that included in their ranks the locally famous Liddells and Montagues. The combined families pooled their resources in order to minimise the effects of financial stumbling blocks such as wayleave rights. They also shared the entire infrastructure associated with extraction and transport of coal, including costs associated with the construction of new and better wagon-ways. The downside of all this – at least to those excluded from the brotherhood – was that the Allies formed a 'closed shop', going out of their way to prevent outsiders from sharing the largesse. They bought up pits and wayleaves and acquired potentially useful stretches of the river bank with no intention of ever developing them; just to prevent their acquisition by the opposition. Consequently, of the nine new collieries, opened between 1725 and 1750, eight were in the ownership of the Grand Allies. If they made a lot of enemies along the way they were nevertheless able to provide a living for many miners and their families in the north east including the young George Stephenson. Over the course of the next fifty years they also changed the face of coal mining from what had been little more than a cottage industry to the commercial leviathan it became. They were also responsible for some spectacular achievements; prominent among them being the construction of the Tanfield Wagon-way. Although they couldn't know it at the time this would be their lasting legacy.

Rich coal seams at Tanfield in County Durham outcrop on the eastern slopes of the Pennines. Between the coal mine at Tanfield and the Staiths on the Tyne, however, were a succession of steep sided hills and valleys. This meant that if a connecting wagon-way was built, a whole raft of revolutionary construction techniques was going to be required. The skills they subsequently pioneered have since become standard practice in railway engineering.

The building of the wagon-way from 'Upper Tanfield' included the crossing of the steep sided Beckley Burn Valley, immediately beyond the mine workings. The Grand Allies imaginative solution was to build a single span stone bridge 30 metres long and 25 metres high. Stone bridges had been around since Roman times but none were built to bear the weight of laden coal wagons. It is a credit therefore to the designer that the bridge is still there today, albeit now isolated from the railway of which it was once a key component. Further on, the Tanfield Wagon-way re-crossed the Burn again, but this time the Allies culverted the stream and raised a railway embankment over it using rock hewn from the intervening cuttings. The wagon-way was wide enough to take two parallel sets of rails and was engineered such that the track bed had a slight fall from Tanfield to Tyne. Thus loaded trucks could roll all the way down to the River propelled under their own weight. Only empty wagons therefore needed horses to haul them back to Tanfield. For this the horses used the parallel line on the return journey, having run behind the trucks on the descent to the Tyne. To control the speed of descent, each truck was provided with a crude brake consisting of a block of wood attached to a lever. To control the descent the wagon man jumped on the moving truck to apply the brake. For some reason the brakes became known as 'convoys'. If a number of wagons were coupled together for the descent the term applied to the group of coupled trucks, all of which were under the control of a single brakes-man. This probably gave rise to

The Causey Arch Bridge over Beckley Burn near Tanfield.

the term 'convoy' we understand today. Convoy brakes were crude at best and no use at all on steep inclines. Some better means of braking trains consisting of groups of coupled wagons was needed and towards the end of the century a simple solution was devised. A few years ago I volunteered to help out on a heritage railway and saw the new method at first hand. Essentially a chain connects brake levers between each truck; the chain running the length of the train. Should any wagons start to run away, the brake only needs to be applied at the rear of the train to draw the connecting chain taut and apply the brakes on all the wagons. Similarly a horse harnessed to the rear of the train can control the speed of the train since the brakes automatically apply should the horse slow down. An impact on the rear of the train would have a similar outcome, tightening the chains between wagons and applying the brakes. A slightly more sophisticated form of continuous braking, pre-dating the fail-safe continuous air and vacuum brakes, employed a similar method to control the speed of trains and is the principle behind the guard operated brake van that brought up the rear on freight trains.

In respect of the loading capacity of the wagons there seems to have been little standardisation in the early days. This led to various disputes as to whether or not a customer had received his fair share of coal when he purchased a 'wagon load'. The largest of the wagons could take a couple of tons of coal and became known as 'chaldrons'. The word is derived from the Latin word for hot; the link being that coal is used for heating things up. The term had earlier been used to define the loading capacity of keel boats, but it became a standard measure of coal (approximately 2½ tons) once the technology for transporting it in bulk became commonplace. Horse and wagon combinations were known as 'dillies'. The origin of this name is more obscure. Dilly, according to the Oxford English Dictionary is a corruption of the word delightful, but this use in a coal wagon context, seems unlikely. Whatever the source of the term, 'dilly' later reappearsed in a significantly different context.

When the Tanfield wagon-way opened its doors for business in 1725 it was the high speed highway of its day. No longer a simple wooden surfaced road, but a genuine railway in the sense we understand it today. It just lacked iron rails and mechanical motive power to complete the picture.

There has been much argument about the gauge, or width between rails (4 foot 8½ inches), adopted in this country and throughout much of the world, but how it came to be adopted seems more down to luck than good management. At the beginning of the railway age there was no national standard gauge – although a width of 4 to 5 feet was common in the north east. There was nothing original in this particular width between rails. It was, and remains, the minimum that would allow horses enough space to move freely; the same gap apparently applied

Chaldrons at Beamish.

between the wheels of chariots belonging to our Celtic forebears. Its adoption for wagon-ways was therefore inevitable given that the first coal wagons were going to be horse drawn. The specific width of 4 foot 8½ inches is more intriguing. Grooves cut by the carts entering and leaving the stone gateways of the Roman forts in Britain, including those on Hadrian's Wall, were found to be 4 foot 8 inches wide and it has therefore been argued that this was the reason George Stephenson applied this particular gauge to the wagon-ways at Killingworth Colliery; where he was working when he built his first steam locomotive. However, since the wagon-ways were there before Stephenson this theory seems unlikely. A more reasonable assumption is that Stephenson just used the rail width already present at Killingworth. Why Killingworth chose this particular gauge for their wagon-ways is more of a mystery; it probably just suited local circumstances at the colliery. Since the Stephenson gauge from the Killingworth wagon-way was exactly 4 foot 8 inches, it is not clear, however, where our national rail network acquired the extra half inch, although the crude tolerances applied to rails at the time probably rendered such specifics meaningless.

The gauge of the Tanfield Wagon-way as a matter of interest was 4 foot.

What remains of the railway at Tanfield still impresses. A few miles of the former wagon-way became part of our national rail network and is now a heritage railway with an eponymous name. Since the line never completely closed, even during the terrible decimations inflicted by Dr Beeching, it lays claim to being the oldest continuously operated railway in the world. Causey Arch is still impressive; in 1725 it must have been one of the wonders of the age. Tomlinson in his book *North Eastern Railway* relates that an 'eminent antiquary' of the time, Dr William Stukeley, who was on holiday in Newcastle, made a special detour just to see the wagon-way. His astonishment is clear: 'valleys filled with earth, 100 foot high, 300 foot broad at bottom: other valleys as large have a stone bridge laid across: in other places hills are cut through for half a mile together...'

Since the wagon-way was so expensive to build it must have been a blow to the proprietors when Causey pit caught fire and burned out just fourteen years after it opened.

So why did the Grand Allies go to all this trouble to build a wagon-way? In the rest of the country, arrangements were already being made to connect the various centres of commerce with canals; surely a more energy efficient transport system for shifting coal in bulk. The answer to this lies in the particular topography of Durham and Northumberland. Little of either county is flat and the three major rivers are shallow except in their short estuaries. Canal construction from the central Durham coalfields would therefore have been very costly. Even a hundred years later, when canals had become the principal means of transporting bulk goods in the UK, they were still non-existent in the north east. Indeed it was because the idea of a canal connecting the Durham collieries to the coast via Stockton & Darlington had been rejected (in 1818), as being too expensive, that the first steam hauled public railway was built.

The Growth of Wagon-ways

Tanfield railway today.

Further developments in railway infrastructure would follow. With the opening of the Tanfield wagon-way we see the appearance of level crossing gates. These, as today, were located at intersections of railway and road although the one at Lobley Hill, was used to define the junction of three separate wagon-ways. In that particular instance the gates acted as a crude traffic control measure, preventing collisions between wagons arriving simultaneously at the rail intersection. Opening and closing of crossing gates was mostly left to miner's widows who received a nominal payment for the service, a sort of company charity since the women had no other source of income. By the same token the widows were provided with accommodation near to the crossing gate; an early form of 'welfare' with mutual benefit to both mine owner and widow.

The political situation in the country during the eighteenth century was febrile at best. Three invasion attempts were made to reinstate the ousted Stewart dynasty. These and the subsequent Napoleonic wars against the French not to mention the revolutionary war in America led to fundamental changes in iron manufacture and a dramatic increase in coal mining. In little over a hundred years coal production across the UK expanded from 2½ million to 10 million tons. Since iron was needed for both arms manufacture and shipbuilding, the smelting process soon exhausted the ready supply of wood. The natural alternative fuel was coal and vast deposits of coal were locally available but coal had already been tried in the smelting process with only limited success. This was because impurities such as sulphur made iron too brittle. It required a number of technical innovations before coal was the first choice as fuel for foundries. Fortunately progress was being made.

Firstly Abraham Darby, in Coalbrookdale in Shropshire, devised a method for removing most of the unwanted impurities by boiling off the unwanted volatile contaminants using a limited supply of air [5]. This produced coke, which it was hoped would make an acceptable substitute for charcoal, however there was an unforeseen side effect. The coking process enriched the concentration of phosphorus in the metal and high concentrations of phosphorus also made the resulting iron brittle. The metal produced by this manufacturing technique proved to be only suitable for casting. Some method was therefore needed whereby, if coke was used instead of charcoal, it wouldn't increase the phosphorus content of the resulting metal. Henry Cort provided the answer. His solution was to keep the fuel and metal apart during the final casting – in effect to melt the cast iron in a ceramic kettle. It must have been annoying that when this was first tried it was found iron still needs a small amount of carbon to keep it malleable. Cort therefore added back just the right amount of coke to the molten metal to ensure its malleability.

With the perfection of this technique there was a rush to buy coal.

CHAPTER THREE

Wylam Colliery & Christopher Blackett

It could be argued that were it not for the devious machinations of Bonnie Prince Charlie the village of Wylam might have had more strategic significance than it does today. At the beginning of the eighteenth century, indeed, its future looked secure. The village lay on the main east to west highway between Newcastle and Carlisle and was the first major crossing point on the river Tyne west of Newcastle. The arrival of the Young Pretender's army at Edinburgh changed all that. As the Stewart army marched on Carlisle, soldiers of the Hanover army, based at Newcastle under General Wade, headed west to confront them. To do this they were obliged to use the only local cross – Pennine highway available to them – the unpaved road through the Tyne Valley skirting the north bank of the river at Wylam. The army was soon, literally, bogged down in steely northern 'clarts' and after two months fruitlessly slogging through mud, Wade cut his losses and headed north to build a new road on more stable ground south of the rocky escarpment that marked the Roman boundary between the barbarian Scots and the 'civilised' British – known as the Whin Sill. The road Wade built sliced through Hadrian's Wall to leave a permanent scar on this remote and beautiful landscape. It is still there today and still called the Military Road; being now notorious for depriving vehicles of their exhaust systems because of its roller coaster undulations. If Wade's road greatly improved lines of communication between the east and west the effect it had on Wylam was to leave it stranded and isolated. As the first fordable stretch of the Tyne the village had begun to expand so that by the beginning of the eighteenth century it already had a cluster of houses, a few shops and several pubs straggling along the northern approach to the ford.

With the new road now siphoning off the traffic, Wylam might have faded away were it not for the discovery of coal. Four coal seams outcrop on the surface here on the north side of the Tyne valley. All are shallow, 3 to 4 feet in thickness, but the coal is top quality and good for making coke. The seams were exposed on the hillside by the action of a stream, the 'burn', which excavated a shallow valley on its tumbling descent to the Tyne. It was in the burn valley that coal mining at Wylam began, albeit initially solely for local domestic use. Since coal was mined during the day, and above ground, the valley, or to use its north eastern name the 'Dene', acquired the name 'Day Hole Dene'.

As the miners burrowed into the hillside coal extraction became more difficult. Not only did the miners now have to shore up the tunnels they created but, during periods of heavy rain, the mines flooded. To alleviate this, the villagers diverted the burn in 1609 so coal could be worked dry. For decades the villagers pursued the 'day holes' deep into the hill side and as a consequence created the first deep pits in the village. Coal mining was still then a cottage industry. It would take the arrival of the Blackett family to commercialise the process; leading to the exploitation of all the local coal seams. The land came into the possession of the Blackett family in 1679; being originally part of the estate of Tynemouth Priory acquired by Henry VIII during the Dissolution of the Monasteries. Prior to the Blackett family's involvement, coal was being laboriously transported on packhorse five miles along the old Carlisle to Newcastle road to the nearest coal staiths, on a bend in the River Tyne at Lemington east of Newburn. A wagon-way from Wylam to Lemington was obviously needed and so, at the same time the Grand Allies were constructing their railway at Tanfield, the current landowner at Wylam, John Blackett, began putting down his own rails along the course of the old road between the village and Lemington. John Blackett's wagon-way opened for business in 1748 and a pit village rapidly rose around the mines. All the new houses in the village were built and owned by the Blackett family and then rented out to their employees. This had advantages and disadvantages to the pitmen. On the plus side they were provided with accommodation close to their place of work but on the downside the pitmen lived by the grace and favour of their employer – as George Stephenson's father would later learn to his cost after getting blinded in an accident involving escaped steam from the pump he was tending. It is not known, although it seems probable, whether or not the Blacketts' also operated the scandalous 'trucking' system so used and abused by mine and mill owners in the nineteenth century. This obscene method of payment involved paying the men's wages, at least in part, with tokens only redeemable at the employer's own shop. Since the employer could then charge what he liked for the essentials of life, and the pitmen were unable to use their tokens elsewhere, the employer had a stranglehold on his employees that ensured their continuing loyalty – if not their respect.

It was John Blackett who took on John Hackworth as the colliery blacksmith at Wylam. The Hackworth family had lived in the village for many years, but on the death of Hackworths' father, the family home was sold to John Blackett and the money distributed among the large Hackworth family. After accepting the blacksmith job Hackworth leased a pit cottage in the village from his employer. This had the fortunate consequence that John's young son Timothy became apprenticed to his father and so learnt the trade from an early age. The rich coal seams and the success of the wagon-way ensured that the Blackett family became wealthy. A stately home was built, called 'Wylam Hall', situated a mile to the north of the main ('Haugh') pithead, and the family fortune looked secure

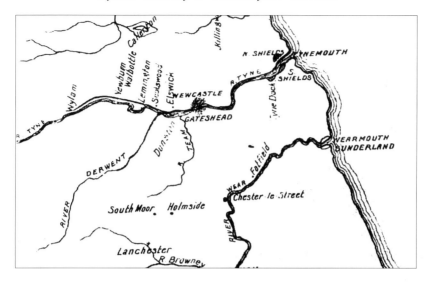

Route taken by the wagon-way from Wylam to Lemington.

until John Blackett died. The running of the mining business then came in to the hands of John's son from his first marriage, Thomas. Thomas had less interest in mining than partying and drank himself into an early grave. Consequently, when his half-brother Christopher inherited the business in 1800, the family mining business was in a bad way. Christopher, though, was made out of sterner stuff than his brother.

Christopher Blackett was born, at Ovingham (pronounced 'Ovinjam') village near Wylam, on 28 June 1751. His mother, Elizabeth Crosbie, was John's second wife. Despite being the son of a mine owner, Christopher had little experience of the mining industry, but had been given a good education by his wealthy parents and possessed the energy, drive and entrepreneurial spirit his half-brother lacked. At the time he inherited the mining business he was acting 'Postmaster' for the City of Newcastle and proprietor of the influential London Newspaper, the *Globe*. The *Globe* was a typical metropolitan newspaper, filled with adverts for guaranteed gout cures and magical hair restorers. It seemed that, like today, men of the nineteenth century drank too much and were obsessed with male pattern baldness. How much control Blackett had over editorial content of the paper is not known but the bulk of the news reported in the paper consisted of military reports on the war with France and Parliamentary pronouncements on the various Pitt progeny. Little news from beyond the capital city was included and virtually nothing from Blackett's native north east. Typical of the paper's parochialism is an entry for 6 November 1805 where, buried under a summary of court proceedings under the heading 'Police', are five lines devoted to a catastrophic explosion in a coal pit 'near Newcastle Upon Tyne' where fifty men were killed including a miner's wife, who died of shock on hearing the news. On the same page is a 100 line advert for 'Ching's Patent Worm Lozenges' – in a newspaper whose proprietor was a north-eastern colliery owner himself.

Christopher Blackett.

Nevertheless, if he failed to promote his birthplace in London then, back home and under his guidance, the development of Wylam colliery, which had stalled under his brother Thomas, now rapidly progressed.

At the opposite end of the country, Richard Trevithick had been conducting experiments with high pressure steam. He had acquired a detailed knowledge of low pressure pumping engines from repairing and rebuilding those installed in tin mines belonging to his father. Mine owners at that time were obliged to pay exorbitant royalties to Boulton and Watt to use their patented process and so Trevithick's motive for experimenting with high pressure steam pumps may have started as a way of circumventing Watt's patent. Trevithick had fortunately received his early engineering training from that advocate of high pressure steam William Murdoch and had probably seen the working model of his high pressure steam locomotive at that time. Unlike Murdoch's employer Watt, Trevithick was enthused with the idea that high pressure steam might be employed to turn the wheels of road vehicles.

In truth Trevithick was an unlikely candidate for the world's first great locomotive engineer. A tall heavily built man he preferred brawling and boozing to scientific and literary pursuits and, whilst well educated, lacked the patience needed to follow a project through to ultimate completion once the thrill of invention had passed. Unlike Watt, he also didn't have Mathew Boulton at his side to develop and promote the abundance of ideas he effortlessly generated. As a result, Trevithick lost out on the revolution that he could be said to have initiated and ended his career ignominiously. A quarter of a century later he had to be repatriated from America by Robert Stephenson, who found him,

Richard Trevithick.

dressed in rags, unable to pay his own fare home, following a typically unwise involvement in a failed mining scheme.

At the turn of the century, however, Trevithick's star was in the ascendancy. Despite his faults, he was unquestionably a genius. If George Stephenson, as Hunter Davies suggests, can reasonably be called the 'Father of Railways', then Trevithick should be considered the 'Father of the Steam Locomotive'. He built his first moving steam driven vehicle in 1801, the year after Christopher Blackett took control at Wylam (and incidentally the year the first public railway, albeit horse powered, opened between Wandsworth and Croydon). His first experimental engine looked to all intents and purposes like the converted stagecoach it was; essentially just a boiler, on a stagecoach chassis, with steam powered pistons connected to the front wheels. As a horseless carriage Trevithick intended it to be used on roads, which is where the initial trials took place. However, the appalling road surfaces of the time, coupled with the carriage's boiler weight, meant its potential as a road vehicle was minimal through no fault of its inventor. As it happened its development never progressed that far. The first evening outing of Trevithick's horseless stagecoach led to the inevitable apocryphal story concerning the operator of a toll gate and the gateman's brief encounter with a fiery demon. Apparently Trevithick had been warned not to use the toll road for fear of damage to the surface, but was allowed to proceed after the gateman. Seeing only flames and hearing the hissing of steam in the dark, he thought he had encountered the devil himself and took flight – a story that constantly reappears in various guises and locations throughout the early days of steam locomotion.

During trials Trevithick lost control of the cumbersome steering mechanism. The vehicle careered out of control, crashed into the side of a house, and toppled over. In typical fashion Trevithick abandoned it where it fell, the fire still burning in the grate, and repaired to the nearest hostelry to drown his sorrows. When the boiler ran dry the vehicle caught fire so by the time an inebriated Trevithick stumbled from the pub nothing useable of his great invention remained. What happened to the damaged house is not recorded. Undeterred by this setback Trevithick built another engine, this time designed to run on rails. He had been supplying his own high pressure steam pumps to other industrial enterprises in the southwest, to the annoyance of Boulton and Watt, and, during a visit to an iron works at Pen-Y-Darren, near Merthyr Tydfil in South Wales, he accepted a wager to build a steam locomotive capable of hauling ten tons of iron nine miles along the local wagon-way.

Essentially the locomotive Trevithick subsequently built was just a variation on high pressure engines he was already supplying for the de-watering of mines excepting that the piston driven wheel, was now connected by rods to wheels on which the boiler was supported. There are suggestions that Trevithick's locomotive was first tried out at Coalbrookdale Iron Works in Shropshire before being sent in 1804, on a sale or return basis, to Pen-Y-Darren. Trevithick, of course, won his bet and used the money to fund further experiments with his steam engine at the South Wales site. Unfortunately it turned out to be the poor quality of the cast iron rails, rather than any problem with the locomotive that would be his undoing. The weight of his travelling engine proved too great for the wagon-way and Trevithick's promising experiments fell apart along with the rails. Nevertheless he was the first to successfully demonstrate the practicality of the idea and proceeded to patent his locomotive design. He just needed the rest of the world now to sit up and take note.

How Blackett heard of Trevithick's experiments isn't clear. It's possible; he might have been told about them via his contacts in the mining industry or it may have been reported to him by his colleagues at the Globe. Whatever the source the timing was perfect. Blackett was by then actively seeking an alternative to the horses being used on the Wylam Wagon-way.

Between 1793 and 1815 the army and navy were engaged in the war against Napoleon; the army, at this time, expanded by a factor of six and the navy nearly tenfold. Horse fodder, because of the war, was in desperately short supply so horses were not only expensive to maintain but also a costly purchase since they were needed in huge numbers by the military. The idea of a mechanical horse, available for work 24 hours a day, that never got ill or lame and, in particular, was fed on unmarketable 'small coal', which Blackett had in abundance stacked about the pithead, was obviously attractive. The Wylam mine owner therefore wrote to Trevithick and asked him if he would build a locomotive for Wylam. Surprisingly perhaps, Trevithick declined the offer citing his ongoing commitments at Pen-y-Darren. Nevertheless, he did agree to provide his engineering drawings for Blackett to use and even supplied one of his own men to supervise the work – a charismatic one legged Geordie called John Steel, who had worked for him in South Wales.

Here the story gets confused. There are unconfirmed reports that the locomotive, later known as the *Newcastle* was built by Trevithick at Pen-y-Darren and then shipped to the north east accompanied by John Steel but the probability is that the engine was built from start to finish on Tyneside; the construction work merely overseen by Trevithick's man Steel. The eye witness accounts that exist certainly support the latter view. These amount to reports from former workers at the foundry that were present at the time the engine was under construction and a letter from Trevithick's widow who recalled the Cornishman visiting Whinfield's foundry 'on at least three occasions' to check on the progress of the work. Given the above, it is reasonable to assume that the *Newcastle* was indeed built at John Whinfield's iron foundry at Pipewellgate in Gateshead. If so, the work took place in the months October 1804 to May 1805. The similarity to the Pen-y-Darren locomotive is striking in the drawings reproduced below:

The *Newcastle* was, in truth, a smaller version of the four wheeled engine Trevithick built for Pen-y-Darren, but made completely out of wrought iron. Like its Welsh forebear it had a boiler connected to a single cylinder that turned an enormous cogged fly wheel. The cogs on the enormous fly wheel intermeshed with cogs around both the front and rear wheels. Cast iron rails were laid down in the foundry to carry out trial runs and the trials went well. There was nothing fundamentally wrong with Trevithick's design but the very man who commissioned the building of the engine then refused to accept it. Christopher Blackett's reasoning was clear. At Pen-y-Darren, Trevithick's locomotive experiments had been abandoned because the weight of the engine had proved too much for the cast iron rails. Blackett knew his own wooden tramway would suffer a similar fate if subjected to the wheels of the 4½ ton *Newcastle*. As a result, the *Newcastle* never went to Wylam; instead it ended its days as a stationary engine at Whinfield's foundry.

Nevertheless, during the period the trials were being conducted at Gateshead, the factory was visited by a succession of names destined to be forever associated with locomotive development. These included the future local locomotive builder Robert Hawthorn and the Wylam quartet – Timothy Hackworth, William Hedley, Nicholas Wood and George Stephenson. Trevithick later even claimed to have visited Stephenson's home during the construction of the *Newcastle*, telling his son Robert, when the latter offered to pay his fair home from America, that he had 'dandled him on his knee' as a child.

Despite his rejection of the *Newcastle*, Blackett hadn't lost interest in steam locomotives and continually monitored Trevithick's progress. Unfortunately, the patience of the Cornish engineer was now running thin. If he couldn't interest the world in his invention then he had better things to occupy his valuable time.

Blackett's *Newcastle*.

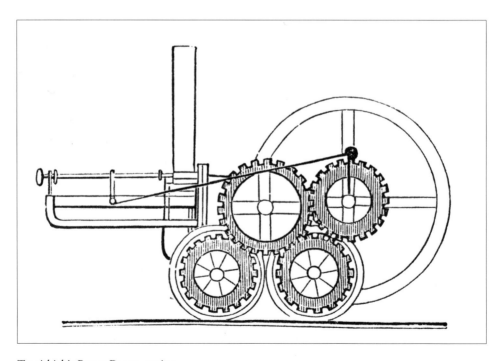

Trevithick's Pen-y-Daren engine.

Nevertheless he made one last effort to promote the concept. He built another locomotive; a much more elegant design, with piston movement transmitted directly to the driving wheels rather than via the meshing of cogged wheels. The new machine had many of the features later locomotive builders would lay claim to having patented. It ran, for example, on smooth wheels, as the majority of locomotives do today and the spent steam from the boiler was fed through a narrow tube into the flow of smoke in the chimney, which created a draught that pulled air through the fire and thus increased the heat supply to the boiler. He called his engine the *Catch-Me-Who-Can* and put it on public display on the site of what is now London's Euston Station. It was meant to arouse both attention and publicity in the metropolis but the whole affair was hampered by looking too much like a fairground ride. The demonstration was held inside a circular fenced enclosure for which a shilling was charged for entry. *Catch-Me-Who-Can*, with one carriage, ran around a circular track and the public were charged a shilling a ride. The public, unfortunately, were as indifferent to the London demonstration as they were to his earlier experiments and, from that time on, a disillusioned Trevithick abandoned his interest in steam locomotives.

This was bad timing as it seemed he now had a market for his travelling engines. Three of the pits at Wylam were returning a good profit – Windsor's Pontop, Simpson's Pontop and Wylam Moor and. Christopher Blackett had by now completely re-laid his wagon-way with plate rails. All he needed was steam locomotives to replace the expensive and inefficient horses he was using. Consequently shortly after the *Catch-Me-Who-Can* débâcle, Blackett wrote to Trevithick asking him to build a steam locomotive but was turned down flat. We can only speculate as to the reason. Perhaps Trevithick felt slighted by Blackett's rejection of *Newcastle* but more likely, given his temperament; he just got bored with the whole thing. He wrote back: 'I have discontinued the business, am engaged in other pursuits and can render no assistance'.

So it seemed Blackett had wasted all the money he'd shelled out improving the wagon-way. As a short term measure he even tried replacing expensive horses with cheap oxen for hauling coal wagons. The oxen even proved capable of shifting heavier loads than the horses could ever manage. However, they were much slower and the experiment was soon abandoned. The only plus point for Blackett was that he got more of a return from selling the animals as meat than he paid for them in the first place.

The preferred fuel for the iron industry at the end of the first decade of the nineteenth century was now coke and in order to meet this demand Blackett built his own coke ovens in the village [7] in 1809. These were the first such ovens to appear on Tyneside and were there to supply coke to a blast furnace recently built at Lemington. However, Blackett can't fail to have noticed that the spoil coming out of his coal mines was a deep red in colour. There was iron ore right here on his doorstep waiting to be won.

CHAPTER FOUR

William Hedley, Timothy Hackworth, Nicholas Wood & George Stephenson

Walking to school in Wylam each day from his home in the village of Newburn, William Hedley passed a small miners cottage next to the wagon-way. Outside the cottage he would have seen children playing, dodging the wheels of the horse drawn wagons rolling by on their way to the coal staiths at Lemington. One of the 'bairns' there may well have been George Stephenson, the second child of the large family that occupied a single downstairs room in the cottage. The two children could little know that their paths would cross many times over the coming years.

William Hedley was born at Newburn on 13 June 1779. His father, also William, was a grocer from Throckrington in Northumberland, sufficiently well-off to pay for his son's education at a small private 'school' in Wylam, in the home of a certain Mr Watkins.

It is a three-mile walk from Newburn to Wylam and if the apocryphal story that Hedley was asthmatic is true, it would have been hard going for the youngster particularly during the winter months when snow covered the ground for many weeks. Despite the rigours of the daily trek, under Watkin's tuition, Hedley blossomed proving particularly strong in mathematics. This led to an early apprenticeship, most likely at Walbottle Colliery, the nearest coal mine to where he lived. By the time he was twenty-one, Hedley had become Walbottle colliery's acting 'Viewer', indicating he was sound managerial material. The resident engineer at Walbottle, at the time, was the future locomotive builder Robert Hawthorn. A friendship developed between the two men, which would stand them in good stead, a few years later when it would be to Hawthorn that Hedley turned for the supply of steam locomotives he needed for a new railway being built in direct competition with the Stockton & Darlington Railway, who at that tim held a controlling interest in the only other existing local locomotive works; that of Robert Stephenson & Co.

In 1803, while at Walbottle, Hedley married Frances Dodds [8] in the same Church in Newburn where he had been baptised. The same pulpit the previous year witnessed the marriage of George Stephenson to his first wife Fanny. Rarely can such a small parish church have had two such influential railway men

Wooden wagon-way.

signing the register in such a short period – although, to be truthful, the barely-literate George could barely manage his own signature at that time. In Hedley's case it was a shotgun wedding. Frances was pregnant at the time with their first child, Oswald, who was born just a couple of months after the blushing bride walked down the aisle. Frances seems to have brought money to the table. Soon after the wedding Hedley began investing; firstly purchasing two small collier boats [9] then taking shares in a lead mine at Blaghill near Alston where he also agreed to act as the mine's general manager. He must have been well thought of in the short time he was there, since when he left, he was presented with a tea and coffee service made entirely out of silver extracted from the mine. Two years after the wedding, a second son, Thomas, was born; the same year the family moved to Wylam. Two more sons, William and George, were born in the next three years – all at the home the family leased from Christopher Blackett, located on the main road through the village and known as 'The White House'. It was a worrying time for a young man with a growing family. There was a desperate need for young fit adults to join the army and fight in the war against France. Even back in England the government had set up a home guard of 'militia' men press ganged ad hoc from the civilian population. George Stephenson was one of those enlisted to serve in 1808 and if he had been unable to raise the cash to pay for a substitute to take his place the course of railway history might well have changed. As it turned out it was the weedy asthmatic Hedley, rather than the burly athletic Stephenson, who, for ten years, served in the Percy Tenantry Cavalry – wasting a deal of valuable engine building time doing military training. Hedley was an unlikely candidate for military service. A portrait of him, made

William Hedley, aged twenty-nine.

Timothy Hackworth.

the same year that the attempt was made to recruit Stephenson, shows a portly figure with an already receding forehead. Given that he was possibly asthmatic it provides a salutory indication of the calibre of the home guard being made ready to defend our nation should Napoleon have crossed the channel.

The reason Hedley moved to Wylam was because Christopher Blackett, fresh from his locomotive experiments, offered Hedley the viewer's job at Wylam Colliery. Since he was similarly employed at Walbottle one assumes Hedley negotiated a substantial increase in salary. In any case it wasn't much later before he purchased his first sea going collier boat suggesting he was now comparatively well off.

As we have seen the evidence of Pen-y-Darren and the damage to the test rails done by the locomotive built for Blackett at Whinfield's foundry convinced Hedley's employer that the wooden wagon-way needed to be strengthened before steam locomotives could be used at Wylam. Rather than replace the wooden wagon-way with iron rails, however, Blackett chose the cheaper option of overlaying it with iron plate. This was Hedley's preoccupation for the following three years. Only when the rails were completely overhauled was Blackett able to turn his attention once again to steam traction. In undertaking the work Hedley was given the assistance of the colliery engine-wright, Jonathon Forster and a young works blacksmith, Timothy Hackworth.

Hackworth was born and educated at Wylam. His father John, a local man, had worked at the colliery for many years and been appointed foreman blacksmith in 1800. Timothy was born in the village on the 22 December 1786 and educated at a local school like Hedley although it is unlikely, given their different financial circumstances, the same school was involved. More probably it was a small self-help enterprise provided by educated pitmen for the use of their own children since Hackworth's father was known to have taught at such an establishment. Like Hedley, Hackworth was good at mathematics and stayed on to the unusually late age of fourteen, after which he started an apprenticeship under the supervision of his father. Sadly his father died when Hackworth was only sixteen and young Tim became the sole provider for his mother, younger brothers and sisters. As his father before him, Hackworth was deeply religious. Although there was no church in Wylam, Hackworth, being a devout Methodist, joined a small Wesleyan prayer group that met each Sunday for bible readings. In later years Hackworth became a lay minister in the Methodist church, travelling around the north east and holding prayer meetings. Consequently he always kept the Sabbath sacred, an issue that would cause him major problems later. On completion of his apprenticeship in 1807, despite being only twenty-one, he was offered his father's old job as foreman blacksmith. Working alongside him, was Jonathan Forster.

Forster was born a few miles from Wylam in the South Tyne valley and had been made engine-wright at Wylam Colliery in 1809, a position he held for forty years. He entered the world in 1775, six years before the man who became his close friend and confidante, George Stephenson.

Stephenson, by then, had moved out of the village but Wylam was still his spiritual home and he kept in close contact with the friends he'd made there. Much has been written about George's life, usually over-exaggerated to suggest that no-one but George had any real part to play in the birth of railways. As we shall hopefully see this wasn't quite the case, however there is little doubt that George was a fundamental element in the mix and rightfully deserves the acclamation that subsequently came his way. So much has been written about George, there seems little new to add, however, since he is a key player in the Wylam story it is worth reiterating a few important facts.

Born at Wylam on 9 June 1781, and hence of a similar age to William Hedley, George, unlike Hedley, never went to school, since paying for a child's education was a luxury 'Old Bob' Stephenson couldn't afford. Instead George was sent to work from an early age; his first paid employment being for the Blackett family, being given a few pence a day for preventing cows straying on to the wagon-way. It is conceivable he also helped with the opening and closing of the crossing gates next to their cottage. The Stephenson home then, as today, was surrounded by farmland; standing in isolation a mile east of the village beside the wagon-way. The Stephenson family rented just one room in the cottage; the rest being leased out to other mining families. It must have been a blessing to the young family therefore that there was lots of open space around the cottage for the children to play in because in winter, with the wind howling round the chimney and five children crammed into one room, it must have been bedlam.

The cottage is now a heritage museum and has been furnished as it was in George's day with most of the space in the Stephenson room taken up by a large four poster bed of the type shared by George's mother, father and two of the younger children Robert and Eleanor. There was also a pull out bed beneath the four poster used by George and his older brother James, and lastly a crib for baby John. The room now looks clean and homely and, with a coal fire roaring in the hearth, almost cheery. When Stephenson lived there it must have been desperately noisy and overcrowded. Everything, eating, washing, cooking, and toilet related activities had to be conducted in the one room; the same room where childbirth was witnessed almost annually. All but the youngest of the Stephenson children, Ann, were born in this room.

The cottage had been purposely built at the point where the wagon-way meets the old Newcastle to Carlisle road – or 'street', to give it its Anglo Saxon name. The building therefore became known as 'Street House'. Immediately to the east was the crossing gate, which was operated by a gatekeeper living at the gatekeeper who may well have been a miner's widow as at Tanfield; but at Wylam his or her duties included keeping tally on the number of wagons passing by, since, from that point eastwards, the local landowners were able to charge the Blacketts way leave rent for each wagon crossing their land. George, unlike Hedley, was heavily built and physically strong. He is said to have taken part in wrestling matches with other pitmen for recreation and was once reputed to have thrown a sledge hammer over the roof of his house in Killingworth to win a

George Stephenson.

wager. He had a pronounced working class Northumbrian accent, which proved incomprehensible to Southern ears, and in consequence would later be forced to rely on others to argue his case against better spoken and educated adversaries. Fortunately George was a plain speaker and made friends easily, which was a considerable asset throughout the whole of his life. It was as much his humble origins and likeability factor that has ensured his legendary reputation as a locomotive engineer when most of his contemporaries are long since forgotten.

The Wylam pit where George's father was acting fireman was one of the first to get worked out and 'Old Bob' was laid off – in the process being evicted from his home in Street House. George was only eight years old at the time when the family were forced to move away. Fortunately his father was not unemployed for long. He was soon appointed fireman at a newly opened colliery at Dewley Burn, but it would be another thirty years before the paths of Hedley and Stephenson crossed again.

The final player in the saga of Wylam and the birth of railways was not strictly speaking a Wylam man at all, the Wylam Local History Society are keen to point out. Nevertheless Nicholas Wood was born at Daniel Farm less than a mile away from the village, in 1795. The farm is on the south side of the Tyne valley, in the Parish of Prudhoe, where Wood's father was tenant farmer. Wood was given a rudimentary education at a private school in Crawcrook before, aged sixteen, he gained an apprenticeship as a trainee viewer at Killingworth Colliery. Killingworth was in the ownership of our friends the Grand Allies and it was on a recommendation from his father's landlord, Sir Thomas Liddell (later Lord Ravensworth), one of the Allies, that they took him on. It was at Killingworth that Wood became friends with a young man responsible for looking after the pit winding gear (or 'brakesman'); a man who shortly became the engine-wright for the colliery. The 'brakes-man' and Wood's future collaborator was George Stephenson. Wood must have acquired an establishment-acceptable bourgeois accent because he later acted as the mouthpiece for his heavily accented friend and colleague.

All the players were now on the stage: Christopher Blackett the forward thinking entrepreneur; William Hedley the educated organiser; Timothy Hackworth the skilled metal worker and Jonathon Forster the skilled mechanic – and watching with interest from the wings, George Stephenson and Nicholas Wood.

The show could now begin.

CHAPTER FIVE

The First Wylam Experiments

With two hundred years of railway history behind us it seems hard to comprehend the controversy that once raged over whether it is possible to get good wheel traction using smooth wheels on smooth rails. Empirically, however, we can see why the early pioneers were so concerned; after all, one of the main reasons smooth rails were used in the first place was to reduce friction between wheel and rail and therefore minimise resistance to the shifting of heavy loads. In the days when the haulage process simply involved a horse running between the rails this wasn't a problem, but when the pulling power was accomplished by a 6-ton locomotive astride the rails there seemed a conflict of interest. Anyone who has witnessed the driving wheels of a modern steam loco, spinning wildly out of control on a wet day, might appreciate the concern.

The odd thing is that Trevithick's locomotives had already demonstrated that smooth wheeled locomotives were practical, from his own experiments at the Pen-y-Darren colliery and in London with the 'Catch-Me-Who-Can' demonstration. The problem was that Trevithick didn't grasp the concept himself and, in consequence, he went out of his way to include caveats in his patents to counter the self-defeating idea that adhesion of engine to rail was going to be a problem. His patents consequently left open the possibility of using some sort of meshing cog device to keep the locomotive tied to the rails in the way rack-and-pinion mountain railways work today. However this approach is normally unnecessary on anything other than the fiercest of gradients; the reason being that friction is both a function of the smoothness of the contacted surfaces and the pressure applied between them. Consequently, the very weight of a locomotive is usually sufficient, except in very wet or icy conditions, to provide all the necessary adhesion to prevent wheels slipping.

However, in 1811,[10] Christopher Blackett was unconvinced and rather than commit himself to building a smooth wheeled locomotive he later found wouldn't work on his wagon-way, he asked Hedley to conduct some experiments to confirm Trevithick's evidence at Pen-y-Darren. Blackett had sound reasons for being circumspect. If he opted for a cogged wheel system he would then have to replace all the existing rails, a time consuming and financially crippling programme of work, and one which he had only recently completed in terms of the laying of plate rails. On the other hand Trevithick's smooth wheeled locos had hardly proved an unqualified success. It was therefore left to Blackett's viewer to provide the evidence on which such an important decision could be made. To do this Hedley enlisted the help of Hackworth and Forster and between them they built an ingenious device they would use for the trials, which Hedley called the 'test carriage'.

It consisted of a purpose built four wheeled wagon fitted with hand cranks connected to a fly wheel linked by cogs to the wheels (*see* picture and diagram). Four men stood on platforms each side of the 'carriage' and turned crank handles to move the carriage backwards and forwards. A small scale model, now in the Science Museum, was made first and only when this had been shown to work was the actual test carriage built. It can be seen from the picture of the model that the carriage's wheels were not flanged and, given their greater size to existing wagon wheels, must have been specially forged for the experiment. Cranking the handles for hours at a time in a semi-crouch position must have been a painful exercise – even to burly pitmen used to working in cramped conditions.

The experiments were held in total secrecy. Blackett knew that if he got the right result he could steal a march over his competitors, reduce his transport costs and consequently the price of his coal. He wasn't prepared therefore to give anything away that could help his rivals. It is a measure of how secretive the work was conducted that the trials were carried out, not on the colliery premises, but on a piece of specially laid down track in the grounds of Wylam Hall. In keeping with the cloak and dagger approach, the experiments were conducted only after it got dark. The nature of the trials are best explained in Hedley's own words. In a letter to a certain Dr Lardner published in the *Newcastle Courant* newspaper of 10 December 1836 he said:

> The carriage was placed upon the railroad and loaded with different parcels of iron, the weight of which had previously been ascertained. 2, 4, 6 etc. loaded coal wagons were attached to it, the carriage itself was moved by the application of men at the four handles, and in order that the men might not touch the ground a stage was suspended from the carriage at each handle for them to stand upon. I ascertained the proportion between the weight of the experimental carriage and the coal wagons at that point when the wheels of the carriage would surge or turn round without advancing it. The weight of the carriage was varied and the number of wagons also with the same relative result.

Hedley loaded the test carriage and then measured what engine weight would be needed to draw specified numbers of coal wagons, before the wheels started to slip. How many experiments were carried out is not known but Hedley describes the experiment as being 'on a large scale', the outcome being, '...the friction of

The First Wylam Experiments

Drawing of 'test carriage' from *Who Invented the Locomotive Engine*, by Oswald Hedley.

The model of Hedley's test carriage.

the wheels of an engine carriage upon the rails was sufficient to enable it to draw a train of loaded coal wagons'.

One assumes Hedley always used the same men placed in the same position on the test carriage on each occasion otherwise there was another variable of weight and physical strength not taken into account. It must have been back breaking work for the 'crankers'. You can imagine the portly and asthmatic Hedley calling out 'Just one more lads.' whenever the pitmen thought they were finished for the day. Once Blackett was happy with the preliminary results from the trials at Wylam Hall, the carriage was moved to the actual wagon-way and exhaustively tested over all the most uneven and steep sections of rail. It was even given a couple of outings along the whole length of the wagon-way, which must have come as a blow to the operators dragged from their beds at 4 a.m. to provide motive power. The results were not published nor is there a mention of them in the locomotive patent Hedley later applied for; indeed in his patent application Hedley even returned to the redundant concept that a locomotive could be fitted with cogged wheels to prevent wheel slip. Regardless of this, having demonstrated the principle to Blackett's satisfaction, it was left to Hedley to build Wylam's first steam locomotive.

As he had no previous experience to draw on, Hedley enlisted the help of Thomas Waters who had worked on the ill-fated 'Newcastle' at the Whinfield foundry where he was now manager. The 'test carriage' provided the wheel base on which the locomotive would be built. Waters made the boiler, at Gateshead, and this was designed to be heated with a single flue from the rudimentary firebox. The steam produced was fed to a single piston that turned the same cog fly wheels previously employed in the 'test carriage' experiments. The other parts of the engine, including the piston and linkage gear were made in the blacksmith's shop at the colliery. No drawings exist of the locomotive, but Hackworth's grandson, Robert Young, suggests it was little different in appearance to the *Newcastle* – which makes sense given Thomas Waters' direct involvement in its construction, but seems improbable if, as is believed, the test carriage provided the basis for its structure. Since Hedley's later engines also used a similar cog driven wheel system to the test carriage it seems probable that the first Wylam loco resembled *Puffing Billy* more than the *Newcastle*. Young goes even further, stating categorically that Waters worked directly with Hackworth in its construction with no involvement from Hedley. This claim should be treated with caution. It was refuted by one of the men who worked there at the time. In an account of one of the pitmen, W. E. Burn, Hedley directed proceedings throughout, providing Hackworth and Forster with hand drawn sketches of exactly what he wanted. The instructions sometimes consisted of brief notes written with a quill pen and posted outside the forge; but more often Hedley just left a few chalk marks on the blacksmith shop door to indicate what changes he wanted. The only tools the men had were a small hand lathe, some hammers, chisels and files. Nevertheless despite its Heath Robinson origins, late in the year 1811, the first of the Wylam locomotives took to the rails.

It acquired the name *Grasshopper*, presumably because of its ungainly stop-start method of moving, and was given its first outing on the stretch of wagon-way between the Haugh Pit, in the centre of the village, and Street House. Hauling five chaldrons of coal the *Grasshopper* hiccupped along the wagon-way, with frequent stops to catch its breath. On reaching Street House, and perhaps sensing Stephenson's symbolic presence, it refused to budge. The boiler, itself little more than a crude barrel made from welded metal strips, had been provided with a primitive form of safety valve consisting of a weighted cover placed over a pipe feeding into it. As the watching audience back off in alarm, Waters loaded additional weights on to the safety valve and ordered the terrified fireman to stoke up the fire. Steam pressure, as anticipated, increased dramatically but, instead of exploding – as the crude boiler had every right to do – it stayed intact and with a shudder the locomotive groaned, sighed and started away again, eventually limping back to the colliery. If hardly an unqualified success, *Grasshopper* demonstrated that the basic principles were sound. Even the enthusiastic Hedley, however, was forced to concede that the engine had fundamental flaws, not least of which was the loss of steam at crucial moments.

For the next twelve months a number of improvements were made in order to improve performance yet the simple modification that might have saved the loco, namely directing the spent steam through the chimney, and thereby increasing the draught through the fire, was not attempted. There was a sound reason for this. Heat loss from the firebox was already so great, without further fanning the flames, that the chimney glowed red hot. According to one observer, 'when the engine was puffing and snorting away, this headlight [i.e. the chimney] would shine like a meteor'.

Nevertheless in fits and starts and on a good day it hauled six loaded wagons 6 miles from Wylam to Lemington without mishap. On bad days however, and there were more of those, it broke down *en route* and needed to be hauled back to Wylam using the very horses it was designed to replace. Nevertheless it was never intended to be anything other than an experiment and the success of the subsequent Wylam locomotives leaned heavily on the experience gained from the *Grasshopper*. During its short time on the wagon-way its progress was noted with interest by other colliery owners and the real possibility of using steam locomotives instead of horses was now seriously considered for the first time, particularly as the cost of horse fodder was at an all time high. Amongst the many visitors who came to see the engine working were Robert Hawthorn, the future locomotive builder from Newcastle and George Stephenson whose former home the locomotive now passed every day. Now employed by the old enemy the Grand Allies at Killingworth Stephenson had been tipped off regarding progress at Wylam by his friend Jonathon Forster who helped build the machine. Over the following months Stephenson made a number of visits to inspect the work, perhaps accompanied by his new work colleague Nicholas Wood.

Samuel Smiles suggests that these outings were only to see the follow up to *Grasshopper*, *Puffing Billy*, during its construction, but the evidence suggests otherwise. Hedley had grown increasingly annoyed by Stephenson's presence and barred him from entry to the colliery before work on *Puffing Billy* began. Consequently, none of the improvements that made 'Billy' such a success and *Grasshopper* such a failure were incorporated into the design of Stephenson's early engines. Of these the most important factor he missed out on was the modification made to eliminate steam pressure loss whenever the engine was under heavy load. Since Hedley was the viewer at Wylam, he must have been well aware that Stephenson's visits during the *Grasshopper* trials were taking place yet he took no action at that time to prevent them. Consequently, there is the sneaking suspicion that he wanted the 'Allies' to know about the problems experienced with *Grasshopper*. The Allies could not know that Hedley had already moved on.

As an experiment the *Grasshopper* had served its purpose well and was not scrapped but relegated to a fixed location underground where it happily lived out the rest of its days as a winding engine. The pitmen who travelled on the wagons hauled by the experimental engine, or were sent to rescue it when it broke down, never knew the locomotive as the *Grasshopper*; to them it was the 'Dilly', the iron horse they despised ,which meant just so much extracurricular work. This must have been especially galling given that they were expected to provide their labour free of charge and in their own time; regularly being pulled out of bed at weekends or in the early hours by a watchman berating them to 'Get up to the Dilly'. The good thing, from their perspective, was they saw enough of the trials of the locomotive to dismiss it as just an eccentricity on the part of the management; a 'boys toy' never likely to threaten their livelihoods or provide a viable alternative to the horse in respect of wagon haulage. Unfortunately their first fears were about to be realised.

CHAPTER SIX

The Development of the Steam Locomotive

The 'secret' experiments in wheel to rail adhesion at Wylam had not gone unnoticed. A fellow colliery viewer and 'Geordie', John Blenkinsop, had seen what was happening and thought he could do better. Unconvinced that adhesion between rail and smooth wheels was ever going to be practical he decided to design an engine that didn't rely on simple friction to keep the locomotive on the rails. His solution was to provide a direct linkage between a cogged wheel on the engine and a rack on the rail. Before applying for a patent he discussed the principle with an engineering friend, Matthew Murray, yet another native north easterner, from Stockton-on-Tees, who ran a foundry and engine works at Holbeck in Leeds.

Murray was enthusiastic about the idea and a joint patent was applied for which they obtained in April 1811. An engine was then built to Blenkinsopp's design at Murray's foundry the following year. This was given the odd name *First Mover*, which they later changed to *Prince Regent*. The locomotive had two cylinders connected to cogs linked to a large central cogged wheel whose teeth meshed with a 'toothed rack' laid down on the outer side of the rail. Other than the cogged system the locomotive had the familiar look of Trevithick and Hedley's prototypes. Blenkisopp then took shares in the J. C. Brandling Colliery at Middleton near Leeds. Blenkinsopp was born in Felling in County Durham and the Brandlings, who owned the Middleton colliery, where his former neighbours. Middleton Colliery already had an operational wagon-way, worked by horses since 1758. This ran 3 miles from the pithead to staiths on the River Aire in Leeds City centre. Blenkisopp's first task was to replace the old wooden wagon-way with his patented rack and pinion system. Once this was achieved he could then try out his experimental locomotive. The *First Mover* made its debut in front of '50 spectators' on the 24 June 1812 when it hauled six 3-ton wagons of coal from the colliery into the City at an average speed of 3 miles an hour.

The rack and pinion system was not designed for speed but despite its plodding slowness it performed well enough and further engines were built; the *Salamanca* later that year and the *Wellington* and *Marquis Wellington* the year after. The idea was even sold to other collieries with Kenton, Fawdon and

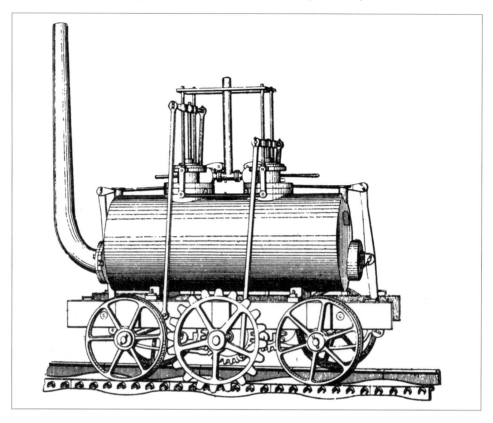

Blenkinsopp's *First Mover*.

Coxlodge in north east England; adopting the Blenkinsopp rack-and-pinion system using engines supplied by Murray from his Leeds foundry of which the *Lord Wellington* is the only named example [11]. Blenkinsopp engines were still working at Middleton in 1828 when they received a visit from the directors of the Liverpool and Manchester Railway (L & MR). The L & MR was then under construction but its owners were unconvinced as to the merits of steam locomotives over horse power, despite the best efforts of their engineer George Stephenson. They must have been impressed by what they saw because, as we all know, it was steam that won out. It is a tribute to the success of the Middleton wagon-way that it is still going strong two hundred years later. Like the Tanfield wagon-way it has become a steam locomotive heritage railway.

While *First Mover* was taking its first tottering steps, two brothers in the north east were attempting something more radical. Stationary steam operated winding engines had been used since the early part of the nineteenth century to haul coal wagons up steep inclines, but William and Edward Chapman decided the same principle could be applied to steam locomotives. Their idea was that a non-moving chain would be connected between two fixed locations. The chain would then pass over a cogged wheel located under the chassis of the engine. The cogged wheel was turned by a piston, as with the Blenkinsopp engines, but as the wheel turned the chain was wound in, and the loco and any attendant trucks carried forward. The brothers obtained a patent for the idea in 1812 and it was tried out at Heaton Colliery near Newcastle the following year. There were problems from the outset. When the locomotive started from a stationary position there was a pause while it took up slack in the chain. It then jolted forward scattering coal and crew in its wake. The Chapmans tried reducing the length of slack by passing it over raised bars, provided every few yards along the track. The chain and cogged wheel were also brought together inside a cowl designed to restrict lateral movement, in the manner of a bicycle chain guard. Unfortunately the chain needed constant lubrication to reduce friction and without constant attention it regularly snapped. After a few fruitless months the idea was abandoned but the brothers at least learnt from their mistakes. They designed a more conventional locomotive, devoid of cumbersome chain drive and this engine, built at the engineering works of Phineas Crowther in 1814, was tried out at Lambton Colliery the following year. It was an odd looking machine, in many ways even more freakish in appearance than its predecessor, however it did have the novelty of being the first loco to use two independent wheel bogies to spread the weight of the boiler and reduce wear and tear on the rails. Lowe suggests in his book *British Steam Locomotive Builders* that an identical locomotive was also built by the Chapmans for Wylam Colliery but there is no other record of this happening and it seems unlikely given the problems the Chapmans were already encountering at Lambton. However, Hedley did alter the design of the Wylam engines around this time, swapping the four wheels for 2x4 wheeled bogies, suggesting that the Chapman brothers may well have had some involvement, even if this only amounted to correspondence with Hedley on the subject.

Of Edward Chapman history has little more to tell, but his brother William became associated with another pioneering locomotive. In 1815, working with an engineer called John Buddle, Chapman built an 0-6-0 for Wallsend Colliery which was given the name *Steam Elephant*. Despite the boiler being supported on six wheels it dismembered the wagon-way and the project was soon abandoned. Little is known about how the 'Elephant' otherwise performed but a reproduction was recently built for the Beamish industrial museum in County Durham, based on the original drawings and a contemporaneous painting of the locomotive at work. The replica at Beamish reputedly performs reasonably well, suggesting Chapman may have finally hit on something. A conventional locomotive, without the cumbersome chain propulsion, its most striking feature is its enormous chimney. The Beamish version, like their version of *Puffing Billy*, possesses design improvements its predecessor could only dream about, including a multi-tube boiler. The jury is therefore out on whether William Chapman finally achieved the success he sought following his failure with the chain system. Either way he now bows out of locomotive pioneering.

These were nevertheless the great early experimental days of steam. Novel methods of steam propulsion appeared year on year; each trying to out-do their

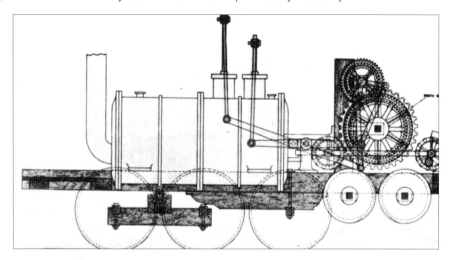

The Chapman brothers' Lambton engine.

Replica *Steam Elephant* at Beamish Industrial Museum.

Puffing Billy at Beamish, above and below.

predecessors by providing some minor variation that would allow the designer to avoid having to pay royalties on existing patents. Of these by far the oddest was the *Mechanical Traveller* designed by William Brunton an engineer from Ayrshire. This was patented the year after the ill-fated Chapman brother's Lambton engine. Peculiar though it was the mechanics of the Brunton vehicle would be well understood by engineers in robotics today. In 1813, however, such a device must have seemed extraordinary – mainly because it walked on legs.

Brunton's father was a clockmaker and the influence is obvious in the design. The single piston was connected to pivoting mechanical legs located at the rear of the engine that pushed the locomotive forward in the manner of someone using a punt on a river. Each leg moved alternately so that at all times there was always at least one point of contact with the ground. In this way the designer hoped that heavy loads could be moved without forsaking any adhesion between wheels and rail. A prototype was built and tried out at Newbottle Colliery where it operated intermittently for about a year with only limited success. Given the numerous stories about the fear generated by early steam locomotives in this instance we can appreciate the concern. It must have been a weird sight indeed as it marched along breathing smoke and fire. We can only speculate what the locals made of it. As it happens their fears, in this instance, were well founded. Towards the end of 1815 the fiery monster exploded, killing and injuring dozens of people. It is interesting that virtually all the failures of early locomotive prototypes, going back to the days of Trevithick, resulted from poor quality materials used for locomotive or rails rather than any fundamental flaw in their design.

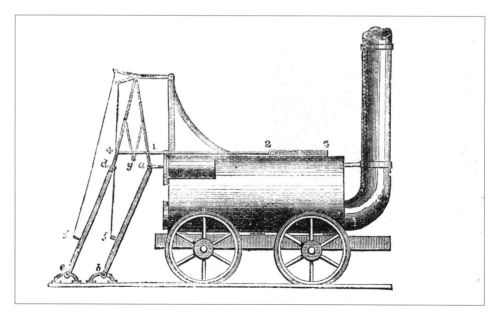

William Brunton's *Mechanical Traveller*.

It might therefore have seemed at the time that fixed winding engines were a better prospect. Little more technology was needed for their introduction than that already established for the pumping engines that had worked so reliably at pitheads for the previous thirty years. Indeed their use for hauling coal wagons along lengths of rail may have seemed the more logical progression. The same sort of fixed engines, already in use for lifting coal out of the depths of the mine or in continuous operation pumping water, would simply be diverted to hauling coal wagons. All that was apparently needed was a long enough continuous chain or rope to connect the mine to the boat staith. There would be a fixed engine at each end of the chain turning a drum over which the chain passed [12]. Nothing could be simpler. Of course the reality was different. Like the failed Chapman experiments, maintaining anything more than a short length of chain or rope was impossible in practice and after a number of failed experiments the use of winding engines for hauling coal trucks was restricted to overcoming short steep inclines where they found their true vocation, as they do today. Standing inclines have the advantage that on downhill sections no extra power was needed. It therefore became standard practice to do away with steam engines completely; using a continuous chain looped over horizontal drums at the top and bottom of inclines in order to move coal wagons up and down steep hills under the effect of gravity alone. Loaded trucks on the descent provided sufficient potential energy to simultaneously haul convoys of empties back up the slope.

It would be fair to say that the Chapman and Brunton experiments in steam locomotion were not the only ones being conducted during this fertile period for original locomotive design, Dendy-Marshall mentions another locomotive called *Iron Horse* built by Taylor Swainson for Whitehaven Colliery, in 1812. Like its contemporaries it seems to have been a Blenkisopp/Trevithick hybrid which failed, in the same way as its predecessors, because the quality of the track didn't match the quality of the engine. This was also the fate of the many novel steam driven road vehicles also appearing at this time. It was only the improvement in the quality of the highway, undertaken by engineers, such as Macadam, which led to the success of the steam traction engine. Nevertheless because so many locomotive trials were being conducted throughout the north east at the same time the rest of the world began to sit up and take note – even if it was then too late to salvage the fortunes of poor old Trevithick.

Driven, by the war on the Continent, to find an adequate alternative to the horse most of the wealthy colliery owners in the north east began experimenting with travelling steam engines at this time, albeit without the glare of publicity associated with the experiments detailed above. Few were successful. In consequence interest in the steam locomotive from onlookers began to dwindle. The world needed a positive demonstration that travelling steam engines would be able to provide a decent return on the considerable financial outlay invested in them. Luckily help was on hand. In 1813 the spotlight once again turned on Wylam.

CHAPTER SEVEN

Puffing Billy

Almost everything we know first hand about the construction of *Puffing Billy* comes from an angry letter from William Hedley to a certain Dr Lardner, which he copied (in part) to the *Newcastle Courant* and *Newcastle Journal* newspapers in 1836. What got up Hedley's nose his having to sit through one of Lardner's lectures in the Newcastle Literary and Philosophical Society in which the good doctor gave George Stephenson sole credit for inventing the steam locomotive and Hedley never a mention. A similar slight to Hedley, would later be made by Samuel Smiles in his biography of Stephenson and challenged by Hedley's son Oliver a few years later. So what does Hedley have to say about the successor to *Grasshopper*?

> Another engine was constructed; the boiler was of malleable iron, the tube containing the fire was enlarged, and in place of passing directly into the chimney it was made to return again through the boiler into the chimney at the same end of the boiler as the fire was...The engine was placed upon four wheels and went well; a short time after it commenced, it regularly drew eight loaded coal wagons after it, at the rate of from four to five miles per hour.

Not the most descriptive or even enthusiastic account so we are fortunate not to have to rely solely on Hedley's description or old engineering drawings to see what Puffing Billy looked like. Both the original locomotive and a working replica exist today.

At the industrial museum at Beamish in County Durham there is a working model of Puffing Billy. It shuffles backwards and forwards along a 100 metres of track pulling a couple of wagons loaded with Museum visitors. In sound and appearance the Beamish model resembles an old Singer sewing machine mounted on wheels. Despite some fundamental differences to the original it is nonetheless easy to see how its forebear operated, at the time in 1813 when it was the world leader. Unlike *Grasshopper* the new locomotive was fitted with two cylinders. These were connected in the same way as the *Grasshopper*, via inelegant rods and cogs, to the wheels, however, *Puffing Billy* had the benefit of

Puffing Billy (as built).

The replica of *Puffing Billy* at Beamish.

a number of improvements. The first and most innovative of these was the boiler. On all previous locomotives water was converted to steam by directing hot gases from the fire through a single flue to the boiler – like the element in an electric kettle. The spent gas was then allowed to escape direct to atmosphere. This was wasteful in terms of heat and expensive in terms of fuel. In the case of *Puffing Billy*, the flue from the firebox ran through the boiler then turned back on itself before the steam was allowed to discharge. Heat exchange efficiency between the fire and the boiler water was therefore effectively doubled.

This innovation, whilst effective, made for a strange look to the engine. As a result of this modification the chimney stack and the firebox were now located at the same end. The locomotive consequently needed two men to operate it; a fireman stoking the fire and a driver facing the direction in which the engine was going to move, on a rudimentary footplate at the other end of the boiler. The other important improvement over *Grasshopper*, not mentioned in Hedley's letter, was that the spent steam was discharged into the chimney through a constricted pipe bent upwards into the flow of smoke. This became, in effect, a crude version of the blast pipe used on present day steam locomotives. The net effect of this high velocity release of steam was to create a draught in the

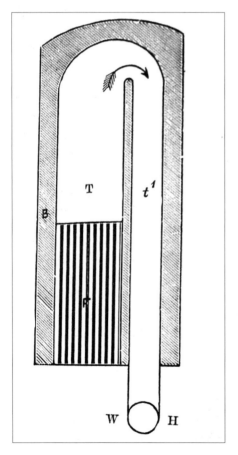

Schematic of Hedley's return flue boiler (b = boiler, T = tube, f = firebox).

chimney and therefore pull more air through the fire. In consequence the fire burned hotter and brighter, generating more steam and a consequent increase in steam pressure. At a stroke Hedley had solved the problem of loss of boiler pressure at crucial moments. Whether he knew exactly what he had discovered remains open to question.

In reality Hedley arranged for steam to escape in this way just to reduce the loud whistling noise that had been the subject of so many complaints during the *Grasshopper* trials. The hiss of steam, discharging directly from the cylinders was reputedly so loud that it caused horses to bolt on the adjacent highway. The ejected steam from *Grasshopper* had also been directed down on to the track which wet the rails and reduced the grip of the wheels. The down side of this modification was that sparks from the chimney, instead of falling on or near the track as previously, were now ejected at high velocity on to the surrounding countryside, setting fire to hedges and hay ricks, for which Blackett was obliged to fork out compensation to recalcitrant landowners. This unwanted side effect, caused by releasing steam through the chimney, should nevertheless have demonstrated, to anyone willing to see, the positive effect that the escaping steam had on increasing the pull of air through the fire. However, if Hedley fully understood the positive benefits of his modification he would doubtless have mentioned it in his letter to Lardner; more importantly it would have been a novel element cited in the patent (No. 3666) taken out for the locomotive in March of 1813. We must therefore assume it was a fortuitous accident – although his son argued to the contrary many years later in his contribution to the 'who invented the blast pipe' debate. Regardless of whether or not Hedley knew what the effect this simple alteration had on raising the temperature of the fire under the boiler was dramatic, its consequences did not escape the attention of another who worked on the engine, Hedley's blacksmith Hackworth, who went on to refine and patent the blast pipe principle, and use it effectively on the engines he later built for the Stockton & Darlington Railway.

Patent application 3666, which Hedley submitted for *Puffing Billy*, is a masterpiece in how to spin nothing out to three pages. Most of page 1 is a sequence of toadying remarks designed to impress both the King and the Patent Office, whilst the remainder says nothing that hadn't already been said better and in more detail by Trevithick ten years earlier. Indeed Hedley's patent is so vague and lacking in novelty it is a wonder it was ever granted [13]. It has even been suggested by Smiles that the patent was actually written by Christopher Blackett and his viewer only acted as 'front man'. This would certainly explain why Blackett paid for Hedley's application himself. Crucially there are no drawings of *Puffing Billy* included or indeed any mention of Hedley's wheel adhesion experiments and their far reaching conclusions. On the contrary the patent misleadingly states:

> ...where the rail is of cast iron with a round or flat surface, I attach or cause teeth or flanges to project from both sides of the wheels of the conducting carriage to

enter the ground between the stones, sleepers or rails so as to prevent the whole from turning or surging on its own axis...

Despite Smiles' claims to the contrary, as far as we know none of Hedley's locomotives ever used cogs for purchase on the rails in the manner of Blenkinsopp engines; indeed the conspicuous lack of these was their most novel feature. The lines about 'teeth and flanges' must therefore only have been included to lead others off the scent or alternatively cover Hedley's backside should his locomotive need to be modified in such a way in future years. This was never a patent designed to enlighten the general public. It seems therefore to have been a smoke and mirrors affair, meant to distract and confuse Blackett's competitors who were watching the goings on at Wylam with ill-disguised interest. The fact that the patent contained no mention of the crude blast pipe or boiler return flue led to much controversy and bitter argument later.

If the blast pipe was a lucky accident then the return flue was a brilliant and original innovation. This perhaps more than any other improvement turned *Puffing Billy* into the first successful conventional steam locomotive. *Puffing Billy*, as originally built, and as seen working at Beamish Museum, had four wheels but, despite weighing less than *Grasshopper*, was still too heavy for the Wylam plate rails which needed to be continually repaired or reinforced. Nevertheless, as a straight replacement for the horse it was a huge success. However, when exactly it made its first appearance is open to speculation. In Young's book about Timothy Hackworth he suggests an August 1813 date for the construction of the 'follow up engine', which he calls the eight-wheeled *Wylam Dilly*. This is almost certainly an error resulting from the common local use of the general term Wylam Dilly as the general term for all the Colliery engines. It does nevertheless suggest a reasonable date for the maiden voyage of *Puffing Billy*. We know that William Chapman advised Hedley on the construction of *Wylam Dilly* in 1814 based on Chapman's own experience with the eight-wheeler he built for Lambton Colliery. We must therefore assume that *Wylam Dilly* was actually built between 1814 and 1815. *Puffing Billy*, the forerunner to *Wylam Dilly*, must consequently pre-date the manufacture of that engine, which again ties in with an 1813 construction date. Regardless of when it appeared *Puffing Billy* was a revelation. According to Hedley's son Oswald, Hedley's engines, 'conveyed sixteen wagons at the rate of upwards of five miles per hour (and) with the carriages empty from six to seven (miles per hour)... doing the work of 16 to 18 horses'.

That it was brought in to replace the colliery's horses was not altogether popular with the working men. Amongst them, after all, were drivers and stablemen, none of whom were going to be needed if the locomotive was a success. Consequently, if the relative failure of *Grasshopper* was viewed with amused disdain by the pitmen the success of *Puffing Billy* was regarded with active hostility. At least some of the subsequent problems with track damage were as much down to sabotage by pitmen, whose future livelihood was under threat, as to the excessive weight of the engine on the rails. These were turbulent years for industry. The country, in 1812, was beset by worker/management conflict. Only a few months earlier a mass trial of Luddites was conducted in the city of York. The accused were brought before the court for attacking and destroying the hated looms of cotton mills in Lancashire and Yorkshire. The outcome was that a number of men were imprisoned or deported; there were even a number of token hangings. In consequence there was an undercurrent of resentment to employers believed to be solely in the business of worker exploitation. The coal mines in the north east were not immune to the national unrest and under cover of darkness the working men took any opportunity available to them to cause problems for the hated 'dilly'.

All hostilities, however, ended on a Friday, the miners' pay day. Dressed to the nines, the pitmen, their wives and children were loaded into coal trucks and taken by *Puffing Billy* to Lemington staith where they climbed aboard a boat, to take them to Newcastle where the women and children went shopping while the men went out on a typical north east bender. The same train collected the human wreckage at the end of the day and brought it back home. It might reasonably be argued, therefore, that the Wylam wagon-way operated the first scheduled locomotive passenger service anywhere in the world. What condition the passengers were in, however, after a night on the town and two 5-mile excursions in a coal wagon doesn't bear thinking about.

The pitmen didn't call the engine *Puffing Billy*, as we have already noted, to them it was the *Dilly*. So where did *Billy*'s name come from? In Smiles' book about George Stephenson, published in 1857, the locomotive is called *Black Billy* so the name by which it is known today wasn't common knowledge near the end of its working life, some forty-four years after it first took to the rails. The operators of the modern day counterpart at Beamish believe the name commonly associated with the engine was originally a derogatory term applied to the Wylam viewer. Since Hedley was overweight and reputedly asthmatic it seems a plausible theory. Only much later did the name of the locomotive and its designer become synonymous. The engine was still not officially known as *Puffing Billy* as late as 1914; being referred to as the *Old Duchess* in Tomlinson's book about the North Eastern Railway. *Old Duchess* may indeed have been its official name; since it could then have acted as a sop to the wife of the Duke of Northumberland who was largely the benefactor bankrolling the Colliery's development. [14]

It wasn't long before *Puffing Billy* began ravaging the wagon-way in the same way Trevithick's engine had at Pen-y-Darren. By now, however, there was a modification that could be applied to overcome the problem, namely the one devised at Lambton Colliery by William Chapman. How much direct involvement Chapman had in the modification of *Puffing Billy* isn't clear, but if Chapman didn't actually build a new engine for Wylam, as has since been suggested, then his expertise was almost certainly called upon when in 1815 Hedley doubled the number of wheels to spread the boiler weight. This may have been a *quid pro quo* for Hedley's assistance with improvements made to the Lambton Colliery locomotive the previous year. Unlike Chapman's engine, which had two separate sets of four wheeled bogies, each pair of wheels on *Puffing Billy* was connected together using cogs linked to a central spindle worked by the two cylinders. This produced additional traction since the locomotive effectively now had an 'eight wheel drive'. As an eight wheeler it continued to operate until the plate rails were replaced with edge rails in the 1820s.

On steep sections of the wagon-way Hedley exploited an idea already used on mines with standing inclines. However, in this instance his locomotive acted in place of a fixed winding engine. A rope was fastened to a post at the top of the incline and connected to a drum on the central cog on the chassis of *Puffing Billy*. The engine then wound itself up the slope using the same principle that winches on 'all terrain' vehicles adopt today. With less load bearing down on the wheels, traction once again became a problem whenever the engine was pulling heavy loads. At such times the best speed *Puffing Billy* could muster was 2 or 3 mph, even with someone walking ahead scattering ash in front of the engine to assist adhesion. The reliability of the locomotive was nevertheless beyond dispute and this attribute became its single most important feature. People came from all over the world to pay court to the *Wylam Dilly*. In 1815 Archdukes Lewis and John of Austria took a boat up the Tyne from Newcastle to Lemington just to watch *Puffing Billy* at work. Its progress must have seemed particularly galling to that other son of Wylam, George Stephenson, whose own experimental engines were operational but performing badly at Killingworth. Ironically the war with France had now ended and the shortage of horses and horse fodder that fuelled the locomotive building experiments was consequently over. Luckily, steam locomotive development had progressed too far to be casually cast aside. The clock couldn't be turned back; steam was in the ascendancy. Its progress would become exponential over the next thirty years.

CHAPTER EIGHT

Stephenson's First Locomotives

Stephenson had now been appointed engine-wright at Killingworth and at a number of other collieries in the ownership of the Grand Allies. As we have already seen, he was watching developments at Wylam closely and was convinced that the future for his employer's wagon-ways lay in replacing the horses currently used for hauling coal wagons with steam locomotives. In typical fashion he proclaimed that anything Wylam could do he could do better. He therefore approached Lord Ravensworth and asked permission to build a travelling engine for Killingworth. He must have made a convincing argument because he began work on it later that year in 1812, just a few months after he was made engine-wright. The 'first' of the Stephenson travelling engines built was called *My Lord* (named after his employer). It is a measure of Stephenson's self-belief that he set out to assemble his loco with no previous experience to call upon just an unlimited amount of self-confidence and a general flare for mechanical engineering. The engine that emerged seems to have been loosely based on Blenkinsopp's cogged wheel *First Mover*, although in the Stephenson engine the cogs were connected by gears directly to the wheels and not to a separate rack running alongside the rails. Although it was reported *My Lord* was capable of hauling up to sixteen loaded wagons at 3 mph it was destined to suffer from the same steam pressure loss issues as Hedley's Grasshopper.

Despite Stephenson's undoubted enthusiasm *My Lord* took him two years to build, not making its full debut until July 1814. It wouldn't take quite so long to build a second engine for Killingworth, which was named the *Blucher* after the Prussian General who had fought at the side of Wellington against Napoleon. The engineering drawing that exists of *Blucher* shows a marginal improvement on the engine that went before. Instead of cog driven wheels there is now a chain connecting the three pairs of wheels, however it is still encumbered by a single flue boiler, and would encounter the same problems of pressure problems that so affected the *Grasshopper* and *My Lord*. Unlike *My Lord* however, *Blucher* had two pistons; one at each end of the boiler. Each of these powered a set of driving wheels and the overall symmetry of the locomotive dramatically improved the balance of the engine on the uneven wagon-way on which it worked. Even in

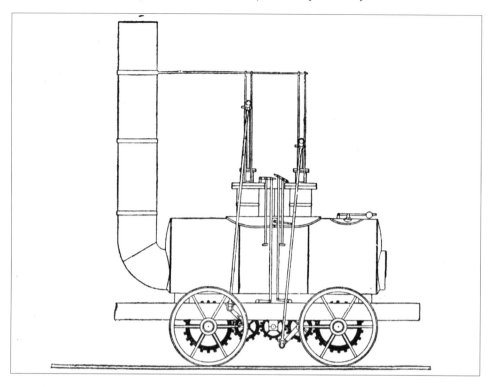

Stephenson's First Locomotive – *My Lord*.

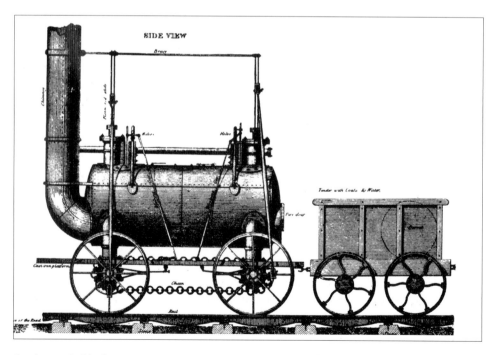

Stephenson's *Blucher*.

Blucher there is still a striking similarity to Blenkinsopp's locomotive in the way the two pistons link to the wheels and we must assume Stephenson also based his design for the second engine mainly on the Middleton Colliery locomotive.

Unlike Hedley, however, Stephenson's locomotive building days were just beginning. Borrowing the Wylam embryonic blast pipe principle he applied it to the *Blucher*, unfortunately improving the performance over its predecessor only marginally, since the expected improvement in increased flow through the chimney was reduced by the large chimney diameter compared with that of the Wylam engines. [15] However, as Blackett had note of, the principal drawback to using steam locomotives on the wagon-way was not the idiosyncrasies of the locomotive itself but the destruction wrought on the fragile rails by the engine's weight. Stephenson's typical radical solution was not to modify his locomotives but to re-invent the rails on which they operated. He therefore invented and patented his own rail, designed specifically to accommodate heavier loads than was then possible using existing alternatives. It could be argued that Stephenson's crusade for better rails was his single greatest contribution to the development of railways in those early years.

Stephenson's great asset was his likeability. He made friends easily and kept them throughout his life. From Jonathon Forster to Samuel Smiles he gathered around himself an abundance of talented people who supported him in the many ventures he embarked upon. Two such talented buddies were Robert Weatherburn and, a relation of Hedley's wife, Ralph Dodds. With their help Stephenson conducted his own track construction experiments at Killingworth and then approached the Walker Ironworks at Newcastle with the idea of making a completely different type of rail. The senior partner there was William Losh, [16] who would again prove a good friend. He and Stephenson would go on to patent a number of improvements to rail and track that found widespread use throughout the country. These included a half lap joint used at the junction of sections of rail to prevent the separate lengths of rail moving independently when a locomotive passed over the joint. Concerned about the pitch and roll of his engines on corners and cambers he also began experimenting with different types of springs on which the weight of the boiler could be supported, patenting his ideas. He was a devotee of stone sleeper blocks, which he used instead of the wooden sleepers commonly in use – his argument being that stone blocks, as well as being more resilient, had the advantage that they allowed horses to run unimpeded between the rails without having to introduce ballast.

While all this was going on Stephenson was also working on his own version of the miner's safety lamp. He had had bitter experience to draw on. In 1806 ten pitmen died in a fire damp blast at West Moor Colliery, where he was the acting brakes-man. Even at Killlingworth explosions were frequent occurrences and his son Robert had had a lucky escape; being blown off his feet while working underground by a detonation caused by fire damp. In 1812 there was an appalling accident involving an explosion at nearby Felling Colliery, which resulted in the death of ninety-six pitmen. Something desperately needed to be done. It is outside the scope of this book to go into the trials and tribulations Stephenson encountered before perfecting his safety lamp however he had the misfortune to come up with the same solution at the same time as the eminent scientist Sir Humphrey Davy who was independently working on the problem and had recently given a lecture on the subject in Newcastle. There followed years of rancorous exchanges over who had discovered the solution first. Davy, being formally educated, had the advantage of including in his patent an explanation of the underlying principle on which both lamps worked – namely the inhibition of flame propagation by the use of a heat conductive wire gauze around the light source – an open candle flame. George, for his part, had to rely on witness testimony to the fact that he successfully demonstrated his own lamp several months before Davy arrived on the scene. In a similar way to the often ludicrous claims made against Shakespeare's authorship of his plays, Davy used the argument that George had neither the formal education nor scientific background to grasp the concepts involved; just one example of the intellectual snobbery that dogged Stephenson throughout his life. The outcome was a score draw. In the end both lamps were used in mines and George received official recognition for his idea and an award of £1,000. The use of the less cumbersome 'Geordy' lamp became widespread throughout the north east and is said led to the name becoming synonymous with people originating in that part of the world.

So how was the semi-literate Stephenson capable of producing technical and, more importantly, legally defensible patents for his inventions? Well once again he had no shortage of friends to assist him. The most important of these, at least in the second decade of the nineteenth century, was Nicholas Wood. Both men came from Wylam and hit it off immediately to the extent as noted earlier that Wood was given responsibility for overseeing the apprenticeship of Stephenson's son Robert at Killingworth. It was almost inevitable therefore that when the Grand Allies gave Stephenson the go ahead to build a locomotive for Killingworth, Stephenson asked the better educated Wood to prepare the technical drawings needed for its construction. The two men would collaborate on the patenting of several of Stephenson's ideas for improvements to steam engines. Wood almost certainly visited Wylam Colliery with Stephenson during the *Grasshopper* trials and drew up the engineering plans for *My Lord* in accordance with the requirements of his Killingworth colleague. Stephenson had by then also received basic tuition and could now passably read and write. Like Hedley and Hackworth, however, he was better at mathematics and his great asset was painstaking experimentation.

He rarely produced the finished article directly from an idea. More often George would get the germ and play around with models until, by trial and error, he came upon something that worked. He was also unafraid of taking risks, both technical and physical. During the testing of his prototype Geordy Lamp he carried out the trials in sections of the Killingworth mine where there were known to be flammable concentrations of methane. At such times he knew for certain his life was on the line. However, not only was George brave he was also lucky. Fatalities were common in the mining industry and, as James Watt had observed, high pressure steam was an unforgiving beast yet the inevitable catastrophic failures of the primitive boilers on his engines somehow never happened while he was around.

George continued trying out new ideas. A third and fourth locomotive was built, each incorporating Stephenson's improvements; not least of which was a first attempt to connect the driving wheels by axles and rods rather than by cogs or chains. In doing so he patented the first ball and socket joints to be used on any locomotive, employed at the junctions of the piston rod ends and the wheels. The later engines also included the first of his patented methods for using springs to correct excessive rolling of the engine. These consisted of standing each corner of the boiler on pistons, which were supported by steam pressure from the boiler. Although in principle a good idea, the unreliability of maintaining boiler pressure using a single flue, which dogged his early locomotive prototypes generally, also reduced the effectiveness of the piston and he devoted a great deal of time to developing a mechanical equivalent. In Stephenson's favour was that he was neither too proud nor too afraid to admit when he got things wrong. A classic example of this was when he recommended a competitor's rail to the proprietors of the Stockton & Darlington Railway – because he knew the alternative was better – even though this meant losing out on royalty payments and a serious fall out with his friend and fellow patentee William Losh.

If George's engines weren't superior to Hedley's they were more widely known because of the many contacts he cultivated throughout the north east, and so it was to Killingworth rather than to Wylam that the great and good began to beat a path for a vision of the future. This was exactly what Stephenson and Wood, who patented the engines, had hoped for and soon clones of the Killingworth locos began to appear on other wagon-ways. George himself oversaw the construction of some of them, even occasionally building the engines himself, including one, in 1818, for the far away Troon to Kilmarnock wagon-way in Scotland.

What George's engines lacked in reliability they more than made up for in availability and the 4-foot-8-inch Killingworth wheel gauge he employed began to be exported along with his locomotives. Over the next seven years Stephenson built two more engines for Killingworth and five for Hetton Colliery. Not only did Stephenson build the locomotives used at Hetton Colliery he also designed and built the wagon-way on which they were to operate. This ran from the mine workings to staiths on the river Wear near Sunderland, a distance of 8 miles. Stephenson's brother was recruited to manage the wagon-way's construction and the line opened for business in 1822. George was a busy man and his services much in demand. Not counting the design and construction of the Hetton wagon-way and the locomotives that were used there he also designed and built three steam driven stationary winding engines for the steep inclines that had to be overcome along the route. Everything George accomplished he did by painstaking trial and error. He always learned from his mistakes, made incremental improvements and moved on. The experience of these early years would prove invaluable as things developed.

Stephenson locomotives at work at Hetton Colliery, 1822.

One of Stephenson's Hetton Colliery locomotives.

CHAPTER NINE

Wylam Dilly & Lady Mary

In a paper presented to Second International Early Railways Conference, John Crompton of the National Museum of Scotland plays down the historical significance of the Wylam locomotives, going so far as to dismiss the idea that there ever was a fourth engine built for Wylam Colliery. Given that he had immediate access to *Wylam Dilly*, on permanent display at the Museum, his views cannot be immediately discounted. It is worth therefore considering his arguments point by point, His principal line of attack seems to be that *Lady Mary* looked too much like *Wylam Dilly* to be a different engine. It does indeed, as to most observers does *Puffing Billy*. This is because *Wylam Dilly* was based on a winning formula Hedley had already established for *Puffing Billy*. It is unlikely that Hedley would choose to change a design he knew worked when he built *Lady Mary*. There are no verified pictures of *Lady Mary* taken during her lifetime, although the railway historian C. F. Dendy Marshall claimed to own a photograph of the engine, which Crompton reproduces in his article. Dendy Marshall had argued that the locomotive shown in his picture, claimed to be *Lady Mary*, displayed too many differences to the other Wylam engines to be just another picture of *Puffing Billy* or *Wylam Dilly*. Crompton, however, pointed out various similarities in the picture to *Puffing Billy*, including an identical ten spoke pair of front wheels. However his similar wheel argument doesn't stand up to much scrutiny. Since *Puffing Billy* started life as a four-wheeler, then became an eight-wheeler, before reverting back to four wheels, there is the probability of wheels (not to mention other components) getting cannibalised and swapped between the Wylam engines. *Lady Mary*'s ten spoke wheels could therefore quite possibly have once belonged to *Puffing Billy* – or *Wylam Dilly* – or *Grasshopper* – or even just cast from the same mould.

More likely, however, is that the photograph actually is of *Puffing Billy*, as Crompton argues – since it shows a four-wheeled, not eight-wheeled, vehicle as *Lady Mary* would have been if and when built. However, if *Lady Mary* was scrapped shortly after its modification to power a boat as I personally believe (see below) it never survived to be photographed and no picture correspondingly exists. There are other arguments which point to the existence of a fourth Wylam engine; for example, the existence of *three*, not two, travelling engines working simultaneously at Wylam was confirmed in a testimony of a former pitman (Jacky Bell) in a letter written to the *Newcastle Courant* on 28 January 1847. There are also numerous references in Archer's book about other locomotives (plural) being built after *Puffing Billy*. Add to this the fact that the renowned railway historians W. W. Tomlinson, and James W. Lowe, writing much closer to the time than Crompton, also include *Lady Mary* in their summary of Hedley's achievements suggesting that in the matter of its existence or non-existence Crompton's is a lone voice. Let us assume therefore, unless contrary evidence emerges, that *Lady Mary* did indeed exist.

If Crompton doesn't believe in the existence of *Lady Mary*, he has no such qualms about its predecessor *Wylam Dilly*, the surviving evidence of which he encounters daily even if he appears sceptical about the importance of the Wylam engines in their contribution to early locomotive development. To the average punter, and I count myself a 'puntee', *Puffing Billy* and *Wylam Dilly* look no different to each other, yet *Wylam Dilly* was indeed different to *Puffing Billy* in some fundamental ways. Unlike *Puffing Billy* it was designed from the outset as an eight-wheeler; the boiler mounted on 2x4-wheeled coupled bogies. One of which was fixed, while the other was allowed to swivel, so the engine could negotiate tight bends, as locomotives do today. We know from William Chapman's correspondence that he advised Hedley on its construction based on his own experience with the two bogie vehicle built for Lambton Colliery [17]. If so, then *Wylam Dilly* must have been assembled sometime after 1814 when Chapman's similar Lambton based engine first appeared. We also know that Timothy Hackworth left the colliery late in 1815 after a dispute with Hedley over extra-curricular work on the 'dilly' so it seems a reasonable guess that by 1815, *Puffing Billy*'s sister engine was already under construction. There was a good reason for setting *Wylam Dilly* on eight wheels; as we know Hedley had plenty of evidence to demonstrate the destructive power of four-wheeled engines on the Wylam plate rails from experience gained from both *Puffing Billy* and *Grasshopper*. In consequence, although Wylam Dilly weighed the same as its predecessor, namely eight tons, its detrimental effect on the wagon-way was much lower so it wouldn't be long before *Puffing Billy* was modified in the same way. The drawing reproduced by Tomlinson, of 'a Hedley locomotive', dated 1815, is almost certainly *Wylam Dilly*. This drawing is often wrongly interpreted as being *Puffing Billy* but Hedley had no specification for *Puffing Billy* – he built it *ad hoc* according to his whims and no engineering drawings were produced. *Wylam Dilly* was a different matter. He knew *Puffing Billy* worked and the specification for the new engine incorporated both the best of *Puffing Billy* and the lessons learnt from its predecessor.

From the days of the first locomotive experiments the two regular drivers/firemen on the engines were John (Jacky) Bell and John Lawson. We know this because of the testimony of Jacky Bell in support of a claim made on behalf of his viewer that Hedley was the true father of the steam locomotive and also

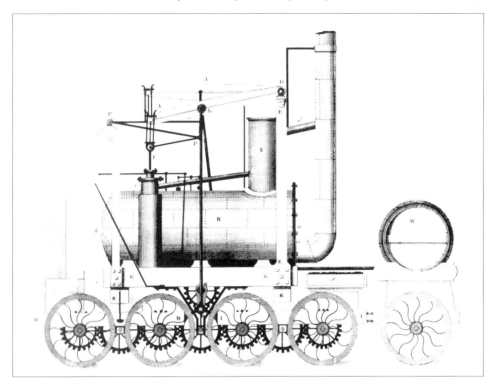

Wylam Dilly (as built with eight wheels).

because his initials and those of fireman John are carved on the firebox of *Wylam Dilly*, now in the National Museum of Scotland. The letter from Bell was published in the *Newcastle Courant* many years after Bell left the pit and how much he actually contributed to it is open to question since he was illiterate and his words were taken down and 'reproduced' by one of Hedley's acolytes. However his letter is interesting in that it attests that he had on occasions driven all four locomotives at Wylam and had no complaints about the reliability of any of them.

Like the earlier engine *Wylam Dilly* generated a lot of interest and may, in a roundabout way, have been the model for the first double bogeyed engines used in the United States, since it was to Timothy Hackworth that the original US engineers went for advice on locomotive construction when he was working for the Stockton & Darlington Railway [18]. Despite the improvement brought about by the modification of the wheel arrangement the 'dillies' continued to damage the thin metal plate rails and Blackett therefore experimented with 'edge' rails on a short purpose built length of track laid down within the colliery yard. The test track consisted of bars of wrought iron 2 inches wide and ¾ inch thick, which were stood on edge and welded to form continuous strips. The experiment was not a success as it was soon discovered that the wooden wheels of the loaded coal wagons were totally unsuitable for the new rails; being unable to cope

Wylam Dilly in the National Museum of Scotland.

with the constant impact of metal on wood. Rather than replace all his wagons, therefore, Blackett unsurprisingly opted to retain the existing plate rails

Locomotive construction at Wylam ended with the building of *Lady Mary*. Why was this? Well with the Battle of Waterloo in 1815, the year *Wylam Dilly* appeared, the war with France ended and there was no longer a need for a cheap alternative to the horse for hauling coal wagons. Similarly, the previously rapacious military appetite for iron also flagged and with it the demand for coal. *Lady Mary* must therefore have been virtually redundant from the moment it rolled out of the workshop. Over the following decade only Stephenson continued to fly the flag for steam locomotives; his farsighted enthusiasm proving crucial in maintaining interest in the concept. If the construction of *Lady Mary* followed closely on that of *Wylam Dilly* it must be assumed her ladyship was built in the years 1816 to 1820; this time without the benefit of input from Timothy Hackworth. The engine's name presents us with difficulties. Who was the Lady Mary whose name the engine acquired? The most likely candidate is Mary Stuart, the daughter of the then Prime Minister, who was married to Lord Lonsdale, an important customer of the Blackett family who lived not far away from Wylam along the Carlisle road.

With the end of the war, unemployment in the country rose significantly as battle weary soldiers returned home looking for work. With so many young and fit men out of work, petty crime rose accordingly; not least in the mining industry where numerous opportunities for scams presented themselves in the tortuous network between the coal face and the customer. It seems to have been a daily concern for the Wylam viewer. The 'Rare Books' section of the British Library possesses a well-thumbed, annotated and signed copy of one of Hedley's books, purchased many years after his death by his ever loyal agent Mark Archer. This book contains several handwritten notes and corrections made by Hedley himself, suggesting it was a constant companion. The book had been commissioned by leading lights in the mining industry of northern England, including Christopher Blackett, and is incorrectly listed in the British Library as having been written by Hedley himself. In fact, the author was one Robert Edington; the proprietor of the Stella Pit. [19] Addressed as a long open letter to the reigning monarch, Edington summarises the various swindles and frauds plaguing the mining industry in 1817. He provided the book with the wonderfully succinct title:

> A Treatise on the Abuses of the Coal trade commencing with the shipping of Coals in the Principal Ports of the North and Proceeding with the Carrying Trade Delivery etc. more especially in the Port of London – the impositions to which the Dealers and Consumers are at present liable with hints and suggestions for the Amelioration and Remedy.

In the tirade that follows, it is not pitmen but distribution middle-men that are bombarded with the most flak, e.g.:

I remember a few years ago many of the coal merchants were respectable but I am sorry to find much of that respectability has dwindled of late, and we now find tailors, tinkers, coblers, bankrupts, attorneys and persons of other professions have deserted their own employments to join the black corps and commerce dealers in black diamonds.

It is interesting that Edington places attorneys alongside bankrupts in his estimation of their worth. With no contemporaneous equivalent of 'Trading Standards' the mining middle-men made hay. Between the loading of a keel boat and its arrival destination it seems that up to a third of the product often went missing or was seriously adulterated. Top quality coal, which brought the highest price in the City, was substituted in transit with inferior material or even just inert ballast. An interesting swindle recounted by Edington concerned two identical coal barges. One was loaded with top quality coal and taken to the London Docks for distribution, but after getting certificated as being all present and correct was replaced during the night by an identical boat loaded with dross. No wonder Edington summarised his own industry thus: 'viewing (the trade) in every stage from the first working in the mine to the last stage of sub-delivery it will be found an unvarying scene of rascality and imposition'.

With Edington's book as his regular muse it is probable Hedley felt surrounded by rogues and vagabonds waiting for the opportunity to steal from him and his employer. As a viewer he had little time for the plight of the working man, which may go some way to explaining his attitude to the keel men's strike of 1822.

The wagon-way provided the means of getting coal to the staiths at lamington, but it was left to the keel boats to transport coal down river to waiting collier vessels at Newcastle. The keel men were therefore in a strong position to dictate terms to the colliery owners and in 1822 they downed tools. There was substance in their grievances; compared to other parts of the country, lighter barge-men in north east England were badly paid for the service they provided and, like mining itself; the work was dirty and dangerous. With no other way of exporting his coal, Blackett once again turned to his viewer to find a solution. Hedley's response was to take one of the locomotives off the wagon-way and adapt it to provide motive power for paddle wheels on a modified coal barge. The conversion cost him £41 and the work was carried out by Hedley's old friend Robert Hawthorn in Newcastle. With the protection of troops to read the Riot Act to dismayed keel men on the river bank the good ship *Dilly* [20] clanked up and down the river towing strings of coal barges all the way to Tynemouth. Each time it strayed too close to the bank or passed under a bridge the blackleg steamboat was pelted with stones. As a result of Hedley's intervention the keel men were eventually ground down and abandoned their struggle; nevertheless the strike lasted a gruelling 10 weeks. It was Jacky Bell who had the un-enviable job of running the gauntlet of physical and verbal hail from keel men on the river bank. The confrontation towards the end became so dangerous that five of His Majesty's war ships were deployed on the Tyne in support of the pit owners;

Keel boat *Dilly* (The *Lady Mary*?).

the most visible display of arms in the north east since the Jacobite wars. One of the warships, moored at Newcastle quayside, fired off a cannon twice a day to warn the keel men, if such a warning was necessary, of the navy's constant presence and vigilance. At the end of the strike the dilly did not immediately return to Wylam. For the next six months it acted as a tug boat piloting ships out to Tynemouth from the City. Hedley's floating *Dilly* has the distinction of being the first steam powered boat ever used on the river Tyne.

The village of Wylam expanded dramatically in the decade 1815 to 1825. Coal deposits directly under the Haugh pit on the north bank of the Tyne had declined but the pithead was retained and the coal face was followed under the river via a connecting tunnel, to be henceforth worked in Nicholas Wood's parish of Prudhoe. Christopher Blackett's son (also Christopher), who served in Wellington's army in Portugal, resigned his commission and moved back to Wylam to assist his father, taking up residence in Oakwood House [21]. An increased demand for lead shot during the French Wars, had also resulted in the establishment of a lead shot foundry right at the heart of the village. This was owned by a company known as Locke, Blackett & Co., but whilst it is tempting to think that the Blackett in the company's name was Christopher Snr, it was more likely one of his cousins Sir Edward Blackett, who owned lead mines nearby. The inner man by this time was also being addressed; local pubs were buying their ale from a new village brewery owned by one William Brown.

CHAPTER TEN

Stephenson, Tennant and the Stockton & Darlington Railway

Despite his earlier attempts to avoid army service, by the December of 1819 Stephenson was training in the Voluntary Reserve along with his good friend and work colleague Nicholas Wood. Given that the wars with France had now ended it begs the question as to why the existence of what we would now call a Home Guard was still considered necessary. Still, as with the end of the First and Second World Wars, there were a lot of fit, weapon trained and disgruntled, demobbed 'squaddies' at home with no employment to occupy their minds and hands. The powers that be were concerned about the possibility of an English equivalent to the French Revolution. The slaughter by the Yeoman Cavalry of unarmed dissidents, seeking parliamentary reform at St Peter's Field, in Manchester (forever known as the Peterloo massacre) had occurred just four months earlier. In a later letter to Joseph Cabery, Stephenson remembered that when army cavalry in full regalia marched through Killingworth, at the time as a show of force, 'reformers' responded by dressing up in mock military uniforms and riding donkeys and parading through the village the following day . These were the days of dissent; riots were taking place right across the UK; many against the Corn Laws which prevented the import of cheap grain from abroad and kept the price of bread, the staple food of the working classes, artificially high. Unskilled men were being replaced in the fields and factories by machines and there was increasing tension between employer and employee, rich and poor, and talk of revolution along the recent French model. George was an unwilling recruit. He was not cut out for the military. We get an impression of his feelings from a contemporaneous letter, 'Mr. Wood makes & (sic) excellent soldier – But I hope we shall never be called to action as I think if any of us be wounded it will be on the Back'.

He was well aware that the action he might be 'called to' would be against the same people he had most in common with; his fellow working man who was being replaced in the workplace by the very labour saving machines he was in the process of building.

Stephenson was a busy man and couldn't really spare time for military service. That he chose to do so suggests that the threat to his beloved machines was more real than any potential events across the channel. How he found time to take on anything new we shall never know. Apart from his ongoing involvement in wagon-ways and steam locomotives, [22] and the battles being waged over ownership of the miners' safety lamp, he was also involved in the construction of stationary engines for Killingworth as well as for other mines outside the ownership of the Grand Allies. Add to these commitments a succession of problems with fires in mine ventilation shafts and you wonder what time he found for other work. Nevertheless, when he saw the opportunity of bringing his beloved locomotives before a wider audience he grabbed the chance.

George was now living at West Moor Cottage near to Killingworth Colliery and was courting an old flame, Elizabeth Hindmarsh. For the years, following the death of George's first wife Fanny in 1805, George and Robert had been looked after by George's sister Eleanor but she left to get married the same year George started militia training, which gave impetus to his courtship. He married Elizabeth the following year. His days as engine-wright at Killingworth were now numbered; a few miles away to the south, momentous events were taking place.

The owners of South Durham collieries were gazing enviously on the wagon-ways of Tyneside. The river Tees was barely navigable beyond Stockton-on-Tees and the nearest productive coal mines in the area lay several miles to the north and west, in the vicinity of Shildon. The idea of building a canal to connect the collieries to the navigable stretch of the river had been considered as early as 1811, but discarded solely on cost issues. However, a wagon-way was another matter. There seemed no financial impediment to building a railway, provided enough wealthy people in Stockton were prepared to commit to it. Unfortunately there seemed little enthusiasm amongst the gentry who were more concerned with protecting their hunting, shooting and fishing interests. Luckily some of the south Durham mines were owned by Quakers.

The Society of Friends included many of the great banking families whose establishments we know today. The more entrepreneurial of their members therefore had a distinct advantage over the general populace because being in the brotherhood meant a universal commitment by Friends to help each other. The Quaker movement counted among its numbers some of the wealthiest people in England including the owners of what would become Barclays and Lloyds Banks, not to mention industrial giants such as Cadburys, Rowntree's and Unilever. A break away ecclesiastical movement, the Quakers dressed plainly, refrained from drinking alcohol and spoke in an already archaic form of English liberally sprinkled with 'thees' and 'thous'. Not being part of the landowning establishment, they had no hang-ups about the maintenance of privilege and no problem with raising the capital investment needed to build a railway from their mines near Shildon to banks of the river Tees. Their problem was that if the railway took a direct line from Stockton to Shildon it wouldn't go anywhere near Darlington and this was where the promoters lived. It was an essential part of their remit, therefore, that the new railway should dog-leg via Darlington even if this meant adding an extra nine miles to the journey. The prime mover behind the enterprise was Edward Pease.

Pease was in his fifties when he suggested a tramway following the route originally proposed for the ill-fated Stockton to Shildon canal. His first proposal for a public railway, using horses to haul coal wagons, was rejected at Westminster but after securing additional financial support he arranged for a new survey of the route and submitted a fresh application to Parliament. The new route avoided all the previously identified trouble spots – mainly land occupied by wealthy and reclusive landowners – and the Parliamentary Bill gained assent in 1820, just months after Stephenson got married. It is a common misconception that the Stockton and Darlington railway was the first public railway in the world. In fact there had been a public railway operating in Surrey, albeit worked by horses, as long ago as 1803 so the principle of such a railway, by the beginning of the third decade of the nineteenth century, was already well established [23]. There was no mention of steam locomotives in the Quakers parliamentary bill, but by the time it opened, the Stockton & Darlington Railway was committed to this form of haulage. So what caused them to change their minds?

There are several versions of the historic meeting between George Stephenson and Edward Pease in April 1821, none of which need be taken at face value. Perhaps the least reliable is that of George's biographer Samuel Smiles who suggested that George, accompanied by the educated Nicholas Wood turned up out of the blue on Pease's door step having walked all the way from Stockton. Pease, according to Smiles, was won over by George's bluff northern manner and simple honesty. More likely is that, as Stephenson himself stated, he made an appointment to see Pease, and just took Wood along to provide both technical support and acceptable gentrified demeanour. George by then was not, however, the humble working man portrayed by Smiles. By the standards of the time he was comparatively wealthy. He had recently been awarded a £1000 prize for inventing the safety lamp, was the chief engineer for Killingworth on a salary of £200/annum, had recently been paid for surveying the Hetton Wagon-way and was getting royalties on a number of patents. Not bad when the daily wage for a pitman was a few pence a day. It was therefore hardly a ragged illiterate miner armed only with winning honesty and rural charm that knocked on Pease's door that day. George had actually come prepared with firm proposals for the construction of the railway. He even offered to re-survey the route to reduce the construction costs by minimising obstacles along the way that required significant engineering. He also came with a personal knowledge of and passion for steam locomotives and more or less demanded that Pease come and judge the potential of steam locomotives for himself, by seeing them at work at Killingworth and Hetton. The meeting between the men took place just days before the Royal assent was given to the Railway Bill on 19 April. Pease, before the month ended, offered Stephenson the job of engineer on the new railway.

With Stephenson's prior commitments George could only find time to survey the line and appoint sub-contractors to carry out the construction work. That his employers at Killingworth were prepared to release him even, for that says much about his persuasive character. One extraordinary aspect of the historic meeting related by Smiles, however, was true. The meeting went on so long that Stephenson and Wood missed their last coach home and had to walk 18 miles to the Travellers Rest pub at Durham where they had pre-booked overnight lodgings. Three miles from the pub Wood collapsed with exhaustion; something Stephenson never let him forget.

Smiles states that Pease visited Killingworth the following year to look at the working engines. By then the survey of the railway, conducted by George, with Robert acting as chain-man, was completed and construction work was underway. Since the level of engineering was that of a true railway and not that of the horse drawn tramway originally proposed there seems to have been a tacit understanding that steam locomotives would be used long before construction work commenced. It is unlikely therefore that the Directors of the S & DR awaited the outcome of Pease's Killingworth visit before making their decision; more probably they took on board a fulsome letter of praise from one of the first advocates of a steam locomotive-based national railway network, William James, who visited the colliery shortly after the railway Bill went through. Nevertheless, Pease was provided with a personal display of the engine's best tricks. One of Stephenson's locos was brought up to West Moor Cottage coupled to a string of loaded coal wagons and Pease was given a footplate ride. This was his Corinthian moment. From then on he was convinced that steam was the future. Within a few months he raised the finance for a locomotive manufactory in Newcastle to be fronted by Stephenson's son. Pease also amended his Parliamentary Bill to allow steam locos to be used on the S & DR.

George had little experience of building railways but then again who had. Over the coming years he would learn how to build bridges strong enough to support train loads of coal wagons, oversee a belligerent army of Lincolnshire navvies that created the cuttings and the embankments, and arrange for the manufacture of new locomotives to work the rails he had patented and manufactured. Since he was already receiving royalties for the cast iron rails he invented in conjunction with William Losh, it might be expected he would promote their use on the S & DR but as we know, and to his lasting credit, he chose not to do so – to the annoyance of his partner. Instead he recommended novel wrought iron rails from Birkinshaw's foundry at Morpeth. To appease Losh, in the end, both types of rail were used on the new railway although the Losh cast iron rails were quickly replaced. Not surprisingly the gauge he chose was 4 foot 8 inches, the same as Killingworth. This wasn't an ego trip on George's part, there were good practical reasons for using it. There was a lot of earth to shift during the railway's construction and the S & DR would be able to borrow some of the coal wagons from Killingworth, without modification, immediately the rails were laid. Also, both the Hetton and Killingworth engines were built with wheels designed for running on this rail width and little change, if any, was needed to either the engineering drawings for new locos or the tools necessary for their manufacture. The first rails were laid at Stockton on 22 January 1822 and the railway opened doors for business on 27 September 1825.

Pease had purchased two locomotives from Robert's factory in Forth Road, Newcastle but by the opening date, the works had only finished building one. This was the engine originally called the *Active*, but later known as Locomotion No. 1. [24] Preceded by a man on horseback waving a red flag, the first train out of Shildon consisted of six wagons of coal, a purpose built coach called *Experiment* containing the Railway's Directors, twenty-one wagons fitted with seats and packed with some 300 or so passengers, and, at the tail end, a further six loaded coal trucks. The six coal wagons at the rear were uncoupled at Darlington where the coal was distributed to the town's poor. The train, having now acquired a brass band, then departed for Stockton. By the time they arrived the numbers travelling on the train had reached 650, not counting those clinging to the sides of wagons for short rides. The whole journey took three hours with the sole casualty a navvy who was hanging from the side of one of the wagons who lost his footing, slipped under the wheels and crushed his leg.

Smiles reported that 'the arrival at Stockton excited a deep interest and admiration'. Not with all the residents, however. Christopher Tennant was less enthusiastic. Tennant was born in the Yorkshire market of Yarm in 1781 where his father ran a hat making establishment. He had invested in lime kilns at Thickley near Shildon and had lobbied hard for a tramway or canal to be built direct from South West Durham to Stockton. He was therefore unhappy that the new railway would be diverted, for no good reason, via Darlington. On behalf of the

Christopher Tennant.

business men of Stockton therefore, and using his own money, he commissioned a survey for a canal in 1818. Not surprisingly, the route chosen by Tennant met vigorous opposition from the Quaker financed consortium – an antipathy that followed Tennant, thwarting his plans, for the rest of his life. Nevertheless, the local enthusiasm generated by Tennant's proposal re-energised the interest of Society of Friends who quickly switched attention to the potentially cheaper idea of a railway; following roughly the line originally chosen for the canal.

With the opening of the S & DR Tennant's plans were temporarily shelved; indeed he became one of the new railway's first customers. However, it still annoyed him that lime from his kilns had to undergo an 8-mile detour just to appease the Pease family. The crunch came when the S & DR unveiled plans to extend their railway to what became the town of Middlesbrough, 5 miles closer to Teesmouth, where the river was naturally deeper and easier to navigate. This effectively side-lined Stockton where Tennant now lived. However, he was already contemplating his next move.

So before the first train led by Locomotion No. 1 chugged into Stockton, Tennant obtained the finance for another railway to connect his lime kilns and the south-west Durham coalfields at Willington with new docks he planned to build at Haverton Hill, on the north bank of the Tees, east of Stockton. As a sop to the Quakers he even offered to build his railway jointly with the S & DR, but using Haverton Hill rather than Middlesbrough as the preferred port. The 'Friends', however, turned him down flat and so he pushed ahead independently with his new scheme.

Initially prospects looked good, but with the backing of Tyneside coal merchants and local landed gentry, in particular Lord Londonderry whose land at Wynyard the railway would pass, the S & DR succeeded in having Tennant's Railway Bill thrown out. Tennant was not, however, this time about to be dissuaded and he revised the proposal radically, this time shortening the line so that instead of going all the way to the collieries of Weardale, it terminated right in the heart of S & DR territory, just north of Darlington – a ploy guaranteed to endear him even less to the Darlington Friends.

The new railway, linking what would became Port Clarence (on the north bank of the Tees) to the S & DR line at Simpasture Farm (now part of Newton Aycliffe), was called the Clarence Railway (CR) as a genuflection to the Duke of Clarence, the future William IV. It gained parliamentary approval on 23 May 1828 with Tennant as the majority shareholder in the joint stock company, almost the only instance of him actually holding shares in any of the various companies he instigated or developed.

Tennant was now in the position of siphoning trade directly away from the S & DR and the Quakers were up in arms. According to Joseph Pease, 'War...open or concealed, had been declared'.

And war it was going to be; a war that in their current position the Clarence Railway couldn't possibly win. The problem was that they needed to use S & DR rails beyond Simpasture Farm and, from the outset, the 'Friends' made things as difficult as possible for their unwanted customer. They denied access to financial support from any Quaker-held banks, at a time that funds were desperately needed to build the railway. When the construction work was finished they immediately increased charges levied on CR wagons using the stretch of track they controlled between West Auckland and Shildon. They also insisted that any CR wagons use one of their weighbridges before using their short stretch of line. Needless to say this rule didn't apply to their own wagons, which were just counted past a checker. The intention was obviously to slow down CR trains and nullify the shorter distance they had to cover to get to the waiting collier boats. They also refused to allow horse drawn wagons, owned by Tennant's railway, to use their rails during daylight hours. As a result of their machinations the CR only began to return a dividend to their shareholders after they had built their own independent (Byers Green) branch line to the coal fields and were no longer obliged to obtain running powers over S & DR rails. However, even during the worst of the S & DR excesses, Tennant received the support of a staunch ally; someone equally incensed at having to pay over the odds for moving his coal over S & DR, metals from collieries he owned near Shildon. That ally was William Hedley.

CHAPTER ELEVEN

The Wylam Public Railways

In 1826 Hedley leased Black Callerton Colliery, where Stephenson had once worked, and the following year left Wylam for good; his position as colliery viewer at Wylam was taken by James Gray. Despite seemingly severing all ties he nonetheless maintained an interest in proceedings at the colliery, to the extent that nine years later it would be Hedley who advertised in the *Newcastle Courant* for an under-viewer for Wylam, by which time he had long since decamped to his new home called 'Burnhopeside', near Lanchester in County Durham. After a short spell at Black Callerton he moved to Shield Row, in what is now the town of Stanley and became a partner in Low Moor Colliery, which he purchased from the 'auld enemy' the Grand Allies.

Still an innovator he turned round the declining fortunes of the mine; first replacing the old horse drawn wagon-way with a standing incline provided with a steam driven winding engine built to his own design. He then turned his attention to the steam pumps used to remove flood water from the pits, and designed and built a two-stage pump that proved twice as efficient as the Watt pumps previously employed. He was also still interested in steam locomotion though his own pioneering days were long since gone. In 1835, aged fifty-six, he made one last attempt to build a locomotive starting from scratch. Ignoring the results of his earlier experiments in smooth wheel adhesion he started making a 'Blenkisopp' type rack and pinion engine, designed to be fitted with cogged wheels and meant to be used solely at Low Moor Colliery. He never completed the work. Locomotive construction slipped further and further down the agenda until he finally abandoned the project altogether, just months after it was conceived. Whether he would have completed the work if he had had a Hackworth or Forster at his side is open to speculation, but that was no longer an option. However, this was the time Christopher Tennant began battling with the S & DR to develop his own railway from the South Durham coal field to new coal staiths at Port Clarence. Hedley was, like everyone else, paying premium rates to the S & DR for shifting coal and was happy to lend support to Tennant's proposal; he even travelled to London to give evidence at Westminster on Tennant's behalf. His efforts were rewarded; the Clarence Railway Bill gained Parliamentary assent on 23 May 1828.

William Hedley in his sixties.

Burnhopeside Hall.

As the CR's main competitor the S & DR was antagonistic towards the interloper and, as we have seen, went out of their way to be obstructive – despite the fact that, for a few miles at the western end, the CR were paying over the odds for running rights over a short section of S & DR metal. The use of steam locomotives initially on the CR was not an option. One of the restrictions the S & DR managed to secure in the Bill was the aforesaid stipulation that only horses could be used on their section of line; so it was only when the CR completed the building of their own line, totally independent of the S & DR, that the restriction was overcome. The CR was then left with a problem. The two prominent locomotive manufacturers in the north east were Robert Stephenson and Co. and Timothy Hackworth, based at new works at Shildon. Neither of these engine builders were sympathetic to the Clarence; Stephenson because of his financial and personal commitment to the Darlington 'Friends' and Hackworth because at that time he was contractually bound to build engines for the S & DR. Hedley provided a solution. He supplied the drawings he'd made for the aborted Low Moor locomotive to his former buddy from Walbottle, Robert Hawthorn, who used them to build two conventional smooth wheeled locomotives for the CR. These emerged from his factory at the end of 1835 and consisted of two 0-6-0s built to Hedley's design; the *Tyneside* and, almost inevitably, the *Wylam*. These were the first steam locomotives to work the CR – albeit restricted to freight duties. This would be Hedley's swan song. At the time Stephenson, Hackworth, Hawthorn *et al* were fronting the railway boom, Hedley fades from the pages of railway history. He would have no further involvement in building and designing steam locomotives; nevertheless it is a tribute to the reliability of his engines that the *Wylam* template was adopted by Foster and Rastick at Stourbridge for the design of the first steam locomotives ever to be exported and used in the United States.

One of the last of the *Wylam*-type engines to emerge in the UK was the *Agenoria*, now preserved in the National Railway Museum at York. [25] It was built in 1929 for the Shutt End Colliery in Staffordshire where it would work, in the unobtrusive manner of all Hedley's engines, quietly and efficiently for thirty-five years. The very year it appeared, however, a revolutionary locomotive was launched before the world. It was built by George Stephenson's son at Newcastle and pointed the way locomotive development would hence forth proceed. It was the *Rocket*.

How had Wylam fared in the meantime? Coal, as we know, was being conveyed daily by steam locomotives along the Blackett wagon-way, but passengers wanting to travel between Newcastle and Carlisle, even as late as 1830, still had to use ancient stage coaches that took nine hours to complete the arduous journey over the Pennines. A northern link between the Irish and North Sea was desperately needed and just such a scheme had been proposed in 1825, but encountered so many objections by wealthy landowners that the Parliamentary Bill, which would have allowed compulsory purchase of the necessary land, wasn't sanctioned until May 1829. Among the many Directors

Agenoria (National Railway Museum).

Stephenson's *Rocket* (Science Museum).

of the newly created Newcastle & Carlisle Railway (N & CR) were John Dixon, a committee member of the S & DR, and James Losh who, along with brother William, had co-patented some of George Stephenson's earliest inventions. Stephenson himself was called in for advice on the optimum route the line would take. It followed the Tyne Valley west from Newcastle and crossed the Pennines between Haltwhistle and Wetheral. As with the CR there were a few prominent and influential landowners who fought tooth and claw against its existence and because of their vociferous objections, a clause, similar to that restricting the CR, was included in the Railway Bill, which negated the use of steam engines; in this instance the restrictions even more onerous in that they extended to preventing the use of steam driven stationary winding engines to overcome steep inclines, of which there were many on the originally proposed route.

Restricted to using horse-drawn coaches and wagons, enthusiasm for the scheme waned the result being that eight years after cutting the first sod the only part of the line completed and opened was the short stretch between Newcastle and Blaydon – at that time a quiet village, but now part of the amorphous urban sprawl of Tyneside. It wasn't just the objectors who slowed things down, the terrain didn't help. Some of the greatest railway engineering the world had ever seen would be needed to overcome a succession of formidable obstacles. These included the construction of a 500-foot-high viaduct over the River Eden at Wetheral – an impressive mile long, a hundred feet deep, cutting through the Cowran Hills. If the railway was ever going to be completed, it not only needed more investment, it needed new enthusiasm for the project. It was the intervention of another Wylam son,[26] Nicholas Wood, which got things moving again.

Wood became a shareholder and director of the N & CR in 1832 and when the resident engineer, Francis Giles, walked out of the project, Wood stepped in to act as adviser on the outstanding work. The project had stalled over the lack of commercial investment and Wood urged the Committee to push for a change in the Parliamentary Bill to permit the use of steam locomotives on the line. Given that steam locomotion was by now an established fact in terms of both the Stockton & Darlington and the Liverpool & Manchester Railway, the Newcastle & Carlisle Committee wrongly thought this would be a formality. Consequently, they rashly purchased three steam locos and began running them on the section of the line already opened to Blaydon. Almost immediately, the local gentry objected and the Railway was served with an injunction requiring them to remove the engines from the line and revert to horse power. This produced such a public outcry that an open meeting was organised in Newcastle, chaired by the Mayor, to argue the pros and cons. The public turned out in force and the few present who opposed the introduction of steam were shouted down. Under such local pressure the anti-steam lobby withdrew their objections, the company's financial backers returned, and the way was open for railway construction to continue.

The eastern section of the railway through Wylam opened for business on 15 March 1835. Two locomotives were purchased for the inaugural trip when six

The Wylam Public Railways

North Wylam station, 1951.

Wylam station today.

Wylam station today.

North Wylam station seat.

carriages were divided between two trains; one hauled by a Robert Stephenson engine (the *Rapid*) and the other by a Robert Hawthorn loco (the *Comet*). There were also three bands of musicians on hand; one was employed entertaining the crowds at Blaydon where the procession began, and the other two, taken along for the ride, perched precariously on top of the first coaches in each train. The Mayor of Newcastle and various other local dignitaries were naturally in attendance, however, the late arrival of the Mayor resulted in a late start for the convoy. The throngs of passengers who travelled in and on the trains must have been annoyed by the continual delays along the way. These were occasioned by regular derailments of the more poorly made coaches included solely for the use of the hoi-polloi. How the bands coped with the derailments isn't recorded. The formal opening of the railway was greeted with enthusiasm by all the formerly isolated communities along the way. The *Newcastle Journal* captured the mood of optimism that surrounded the event as follows, 'to describe the manifestations of rejoicing which everywhere presented themselves along the way would occupy a much larger space than we have at our disposal'.

At the scheduled stopping places cannons were fired to greet the arrival of the two trains with the local paper reporting an alarming number of firearms brought out to celebrate the occasion, 'Every villager who could muster a gun, pistol or fowling piece put it into requisition on this interesting and auspicious occasion.'

These ad-hoc gun salutes included a volley from a large crowd of rowdy labourers and farmers who assembled near Eltringham Village and fired off rounds of small arms in unison as the trains went by. There are thankfully no reports of anyone sustaining gunshot injuries on the day. On the return journey from Hexham two of the more decrepit coaches were abandoned and all the passengers, including more than a hundred women, were crammed into the remaining carriages for an uncomfortable, if boisterous, journey home. By this time most of the travellers were too well oiled to care following lavish alcohol fuelled luncheons in Hexham hostelries.

The completed railway finally formally opened up the coast to coast route on 18 June 1838, a decade after the parent company came into existence. Wylam station (or South Wylam as it was later known) was opened in 1835 and continues to serve the local community today.

In 1871 sanction was obtained for another railway comprising a new northern loop off the N & CR, intended to connect Wylam, Newburn and Scotswood. Although there would be additional passenger stations along the route, the line was mainly intended for colliery traffic. The railway was designed to incorporate the whole of the former Blackett wagon-way between Wylam and the Lemington staiths and in retrospect was an odd undertaking. Given the immediate proximity of the N & CR, the potential financial benefit of a separate railway ostensibly serving the same communities was dubious at best – and so it would later prove. The northern loop operated under the name the 'Scotswood, Newburn and Wylam Railway' and included new stations at Wylam (to be known as North Wylam), Heddon-on-the-Wall, Lemington and, Hedley's birthplace, Newburn.

It re-joined the N & CR at Scotswood. The final section, west of Wylam, was completed in March 1876 when a new bridge over the Tyne was declared open. Like the wagon-way it replaced it was initially single tracked, but soon double tracked throughout to accommodate coal trains feeding in from northern spurs from the nearby collieries. Despite moderate returns from coal traffic, it was never the financial success anticipated by its sponsors and was sold in 1882 to the North Eastern Railway (NER) company who had also by then assimilated the N & CR. The loop, as part of the NER, then, perhaps surprisingly, continued to operate right through to the Beeching dismantling of the national railway network in the 1960s, when the line was closed along with all the intervening stations. In this instance, however, there seems more justification for Beeching's closures; how a small village the size of Wylam managed to support two manned railway stations, a couple of hundred yards apart, for more than a hundred years is a mystery.

For a busy thirty year period in the middle of the nineteenth century Wylam became a small boom town, served by two railways and supporting a colliery, a brewery, a lead shot manufactory and an iron works. During this period the population reached a peak not surpassed until the post war Newcastle overspill expansions of the present day. From a perspective of Wylam as the centre of industry, however, all good things had come to an end; resources of iron and coal are unfortunately finite.

CHAPTER TWELVE

Timothy Hackworth and the Locomotive Industry

By the middle of the nineteenth century Timothy Hackworth had carved out his own niche in railway history. After the fall out with Hedley, he went to work as foreman blacksmith for Hedley's former employer at Walbottle Colliery and in the nine years he worked there, showed no interest in locomotive building or maintenance. He must therefore have been surprised when approached by the Pease family and asked to work in a managerial capacity on the Stockton & Darlington Railway. From his Wylam days he had got to know George Stephenson well and it was Stephenson who recommended Hackworth to the Quakers. It is questionable how well the two men actually got on as Hackworth rejected Stephenson's advances several times before finally accepting the offer. What may have eventually convinced him was the opportunity provided to dip a toe into steam locomotive manufacture without fully committing himself, as by this time he was already planning to set up his own engine works.

In the subsequent loan spell at Stephenson's new locomotive works at Forth Street, Newcastle, Hackworth acted as a temporary replacement for George's son Robert who, at Hackworth's suggestion, unwisely took up an offer to work on the nascent railway system in Colombia. Upon Robert's ignominious return to Newcastle, Hackworth happily returned to his old job at Walbottle. The catalyst for a permanent move came with the opening of the S & DR. Timothy was offered the post of Superintendent and given responsibility for all aspects of the running of the new railway. In particular, he was in charge of operation and maintenance of the new locomotives emerging from Forth Street. George Stephenson, never one to refuse any paid assignment was, by then, committed to too many engineering projects; in particular he was acting chief engineer for the Liverpool & Manchester Railway. His expertise and practical ability was consequently spread dangerously thin to the extent that he was a virtual stranger at Forth Street. For a time it seemed that the works at Newcastle might close until Hackworth agreed to step into the breach and was able to oversee the production of Stephenson's locos.

Hackworth had, it seems, little, if any, involvement in the engine's design although he may have offered some suggestions as to how Stephenson's locomotives might be improved. It took four more years before Hackworth was permitted to build his own engines for the S & DR without input from Stephenson. The first to emerge was the *Royal George*, whose appearance couldn't have come at a better time for the Quakers. Stephenson's engines were now so unreliable that the Pease family were thinking of doing away with them and reverting to horsepower, a situation that already existed at Hetton, where all of Stephenson's much lauded locomotives had been taken off the rails and replaced four legged equine equivalents. Stephenson's rapid fall from grace is well illustrated by comments in a letter to Hackworth from Stephenson's friend and colleague, John Dixon written in February 1826:

> I think he (Stephenson) is out of favour with the Committee now; alas how fleeting and transitory are the smiles and support of men. It is not more than 3 years ago that GS was looked upon as more than human and almost deified by Edward Pease and even Mewburn, and Storey was scarcely expected to continue his situation but the tide is turned in a contrary direction.

Stephenson had certainly taken his eye off the ball in his native north-east. For the crucial first three years of the S & DR, his precious time was employed doing engineering work for the Liverpool and Manchester Railway. He had therefore neither the energy nor the enthusiasm to gainsay the gradual replacement of steam locos by horses on Teesside. The situation became critical in the summer of 1828, at which time there were no operational locomotives left anywhere on the railway. The Pease family formally complained to Stephenson and as a stop gap Stephenson provided them with a method for conserving the energy of the horses now deployed throughout the railway network. [27] This consisted of an open ended wagon, which could be connected to the rear of coal trains, and on which a horse could rest on downhill stretches of the line. Stephenson called it a 'dandy cart' and his dandy carts continued to be used right up to the beginning of the twentieth century wherever horses were used for haulage on rail or tram lines.

This didn't help Hackworth. He was desperately trying to keep Stephenson's locomotives rolling. Nothing better illustrates the problem than the growing notoriety of Locomotion No. 1. Following its spectacular appearance on the opening day of the S & DR it proved reliably unreliable, spending most of its days undergoing repairs at the Forth Road workshops. The final straw was when the boiler of the engine exploded at Aycliffe Lane killing the poor driver John Cree.

New light on this and other important aspects of his life at the time have been revealed by the acquisition by the National Railway Museum of a century's worth of correspondence and paperwork accumulated by the Hackworth family; in particular the letters of Hackworth's son John, son-in-law George Young and grandsons Robert Young and Samuel Holmes. Their aim was to place Timothy in the top tier of railway pioneers; however, in so doing they needed to belittle

Dandy Cart (National Railway Museum).

the contribution made by the opposition. These will be discussed later but the Archive contains some interesting insights from the time Hackworth was acting Superintendent of the S & DR. Samuel Smiles acknowledged in his biography of Stephenson that, 'For some years...the principal haulage of the line [the S & DR] was performed by horses'.

He suggests this was because the general slope of the land was towards the sea and therefore horses were the cheapest mode of traction. This was no doubt correct but it has to be remembered that three of Stephenson's locomotives had already been expensively purchased for the railway. Since, in theory, one steam loco could pull fifty times as many loaded trucks as a horse, the fact that horses were still the preferred option for haulage is a good pointer to the condition of Stephenson's engines at the time. The truth is they just weren't good enough. When the boiler of Locomotion No. 1 (or, ironically, the *Active* as it was then called), exploded on 1 July 1828, it was the second such boiler explosion involving this engine that resulted in the loss of life. The S & DR attributed it to operator error since the safety valve was found to have been screwed down by the driver. However, this wasn't entirely the driver's fault. He had to do something to keep the engine moving whenever it stalled due to lack of steam, and increasing pressure by this dubious method had become custom and practice. On this occasion the fireman was hurled 24 metres by the explosion and died later from the burns he received. This catastrophic failure of Locomotion's boiler necessitated the first of three complete re-builds by Hackworth and during the first three years of the S & DR there was more down time than operational time associated with Stephenson's locos. One of the letters, in the Hackworth Archive, from one of the railway's committee members William Kitching, makes a number of telling comments. He starts by discussing the outcome of the Liverpool and Manchester locomotive trials at Rainhill, which the Hackworth family would forever believe to have been fixed (I have left the text as written, complete with grammar and spelling errors):

> I shall be extremely glad to have heard that fair play had been allowed to the different engine makers who were competitors for the premium, from the very cool reception that a Timothy received from forth street Geo shewed very plainly he was afraid of the Shildon production, and well they might as they well know that every Engine which they have yet sent has had great needs of mending, the new one last sent was at work scarcely a week before it was completely comdemd and not fit to be used in its present state, the hand gearing and valves have no controul in working it, when standing without the wagons at Tullys a few days ago, it started by itself when the steam was shut of and all that Jem Stephenson could do was not stop it, run down the branch with such speed that old Jim was crying out for help everyone expecty to see them both dashd to atoms the depots being quite clear of wagons, which would have been case had not the teamers and others thrown blocks &c in the way and fortunaly threw it off, a similar occurrence the following day in going to Stockton as soon as the wagons were unhookd at the top of the runaway goes Maniac defying all the power and skill of her Jockey, old Jem nor could it be stopt untill it arrive near the staiths, had a coach been on the road coming up its passengers would have been in most dangerous stituation – the force pump is nearly useless, having had every day it was at work to fill the boiler with pails at each of the watering places. No fewer than 3 times the lead plug was melted out, this Maniac was a forth Street production and at last was obliged to be Towed up to the Hospital by a real Timothy in front on 6 wheels and actually had 24 wagons in the rear as Guard, it is now at head quarters Shildon – There is at present no ships in the River, – from what I have heard of the doings with you it appears that the engine who had a Booth as the inventor of the copper pipes in the Boiler was without either Judge or Jury to be the winner, it would have been dreadfully galling to the owners had not the favord ones carried the day no matter how often they happend accidents.

The letter says much about both Stephenson's engines and the relationship between Stephenson and Hackworth. It is not clear which engine was the dangerous runaway called the 'Maniac', which nearly killed Stephenson's nephew, but it is obvious that its problems verged on terminal if the lead emergency failsafe plug in the boiler melted each day the loco was operational. The perceived superiority of the 'Timothy' engines from Shildon is obvious. It is also clear that there was some friction between Stephenson's Forth Street Works, where the S & DR engines were built, and Hackworth's Shildon works where they were maintained since 'every Engine which they have yet sent has had great needs of mending', including a brand new Stephenson engine condemned within a week of starting work.

By 1827 there were six Stephenson's engines operating on the S & DR. When one of these, the *Hope* (perhaps the 'Maniac' referred to above), ran out of control at the dockside at Stockton, causing major damage to the dock and major destruction to itself, a replacement engine was sought and for the first time the proprietors of the S & DR considered building it themselves. The task was given to Hackworth and he decided to employ every trick he'd learnt in seventeen years of working with steam locomotives. From his days at Wylam he took the design of the return flue boiler and blast pipe, which he refined to increase the draught and later went on to patent. He also mounted the boiler on six wheels to spread the load. From the experience gained from working on Stephenson's engines he adopted springs to compensate for roll and the coupling of wheels to increase both adhesion and pulling power. As an amalgam of the best features of its predecessors it was expected to work and despite some initial teething troubles, fortunately for everybody, it eventually did.

The surprising thing is that at the time a Wylam-type return flue boiler was being incorporated by Hackworth into his new engine, the *Royal George,* [28] a

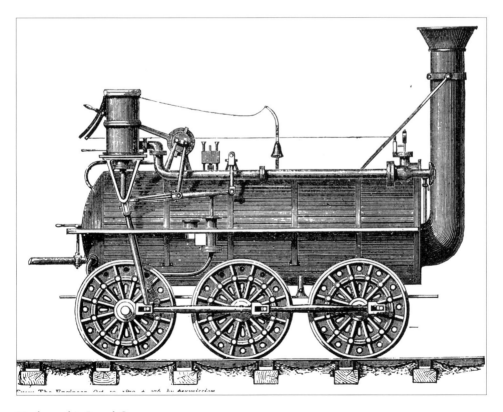

Hackworth's *Royal George*.

modern multi-tube boiler, designed by Robert Stephenson, had already emerged from Forth Street and was being trialled on the S & DR. The truth was it didn't work very well. Robert, freshly returned from South America, was having all sorts of problems stopping the boiler tubes leaking. Why this was a problem isn't clear; multi-tube boilers in themselves were not novel although none had yet been fitted into the boiler of a steam locomotive [29]. Hackworth seems to have been unimpressed by Robert's new engine – the first to be given the name *Rocket* – which rolled badly because of the uneven weight distribution over the wheels caused by the new type of boiler. Robert was already re-thinking the design as the construction of the Liverpool & Manchester Railway neared completion. Much has been written about the Rainhill trials; however to summarise, the Proprietors of the Liverpool & Manchester Railway, being unconvinced as to which method to use for haulage on the line, agreed to offer a prize of £500 to the best device that could meet the criteria set out by the judging panel. This included the requirement for some sort of haulage device costing less than £500 that could pull a load of 20 tons at 10 mph over a fixed length of track. There were dozens of entries and these included the inevitable crop of perpetual motion machines. The majority never left the drawing board and by the final days of the trials there were only four genuine contenders left. These were: a simple boiler on wheels called the *Novelty*, which could nevertheless attain speeds of 40 mph provided it wasn't encumbered with any sort of load; a primitive Wylam type

Hackworth's *Sanspareil* (No.1).

loco called the Perseverance; Stephenson's *Rocket* and Hackworth's *Sanspareil*. Given his recent successes Hackworth was the favourite to take the prize, but he was at a disadvantage compared with the opposition. He had cobbled his engine together in his spare time and, significantly as it transpired, relied on others to provide the main component parts.

Sanspareil's boiler was made at Longridges's Foundry and the pistons came from Hackworth's main rivals, the Stephensons. It was unfortunate therefore, and the subject of numerous subsequent conspiracy theories, not least those promulgated by the Hackworth family themselves [30], that it would be the pistons that failed. In truth *Sanspareil* wasn't ready to 'strut its stuff'. If it had been, it would no doubt have given *Rocket* a run for its money, as in the moments it was working it was the most powerful engine on display. Sadly for Hackworth it broke down too many times and, despite being over the stipulated weight and given second and third chances to redeem itself, it failed to perform to the satisfaction of the judging panel – ironic considering that Hackworth's reputation was built on the reliability of his engines. The fact that the L & MR later purchased *Sanspareil* did little to assuage Hackworth's anger at losing out at Rainhill. He later wrote to the proprietors begging them to reconsider their decision:

> You are doubtless aware that on a recent occasion the Locomotive Engine Sanspareil failed in performing the task assigned to her by the judges – it were now useless to enter into a minute detail of the causes – suffice it to say that neither in material construction nor in principle was the engine deficient, but circumstances over which I could have any control from my peculiar situation compelled me to put that confidence in others which I found in sorrow was too implicitly placed – as the defects were of a nature easily to be remedied – my immediate attention was turned to that point – and I now report to you the extent to which success has attended my efforts. The whole alteration which has been made is the removal of a cylinder which failed from its defective casting.

In short, in Hackworth's view, it was only the faulty Stephenson-built cylinder that let him down. His 'confidence in others' from his perspective had definitely been too 'implicitly placed'.

Despite the problems at Rainhill, Hackworth's confidence in both *Sanspareil* and its predecessor *Royal George* was not misplaced. Both engines enjoyed a long useful working life and it is a tribute to Hackworth's expertise that, after *Royal George*, virtually every engine produced for the S & DR was built to Hackworth's design, including those later made by Robert Stephenson & Co. In total eleven of Hackworth's engines were produced, using the *Royal George* template, between 1827 and 1829 and among their number was the copper chimney, built for speed, *Globe* – the first locomotive designed specifically for passenger work on the S & DR.

Hackworth moved to the new S & DR locomotive works at Shildon in 1826 and stayed there for fourteen years producing over the following decade as series of multi-tube boiler heavy freight locomotives known as the Wilberforce engines. These, according to Joseph Pease, were the best travelling engines ever used on the line. Hackworth terminated his contract with the Pease family in 1840 and took over the locomotive works belonging to his brother, Thomas, on a site nearby at Shildon. The works, built at Timothy's suggestion and partly financed by him, produced locomotives and standing engines for the S & DR even while Timothy was churning them out for the same company at their own works a mile or two up the road. Thomas had also served an apprenticeship at Wylam and was familiar enough with the Wylam locos to add his own postscript to the 'Who invented the blast pipe' debate – unsurprisingly crediting his brother Timothy as sole inventor. The Hackworth-owned factory was named the Soho Works and Timothy stayed there for the rest of his life eventually producing more than 120 locomotives, including several for the London, Brighton and South Coast Railway. He even managed time to build and export the first ever locomotives to be used in Russia and Canada.

Timothy's brother Thomas, who had set up the Soho Works under the name 'Hackworth and Downing' [31], left Shildon to set up his own locomotive manufacturing works in Stockton. He built the first engine to his brother's design for the S & DR therein 1835 – a 0-6-0 called the *Magnet*. The company in Stockton traded under the name Fossick and Hackworth, although, George Fossick, like Thomas' former partner Downing, was solely there to provide the finance. Along with marine engines and freight wagons the firm was soon supplying steam locomotives to the nearby 'Clarence' and Stockton and Hartlepool Railways – yet one more Wylam son making steam locomotives. According to Lowe, Thomas Hackworth went on to build more than a hundred

Fossick and Hackworth 0-6-0 built for the Llanelli Railway & Dock Co. in 1864.

locomotives at the Stockton works between 1839 and 1866, including seven which were supplied to the 'Waterford and Limerick' and 'Dublin and Meath' Railways in Ireland. Locomotive production ceased at Stockton following the death of Thomas' partner Fossick.

Thomas and Timothy remained close throughout their lives, even though they were often competitors in the same field. Timothy was first and foremost a family man. He fathered nine children, three sons and six daughters, although one of the children died in infancy. Much of the correspondence in the Hackworth Archive consequently concerns purely family matters, in particular the anguish and guilt he felt over his daughter Ann's mental illness. Nevertheless, throughout his life, his faith never wavered and he continued to act as a Methodist lay preacher right up to the end, conducting services right across County Durham with regular Sunday sessions at a chapel in Barnard Castle. After his death in 1850 his son John took over the works but by then the railway boom had run its course and the business was sliding slowly into decline. Soho works was sold in 1855 and the site eventually morphed into the great wagon works that operated there until 1984. Hackworth's works now form part of the National Railway Museum with many of Hackworth's original buildings still standing. John Wesley Hackworth became his father's principal advocate, arguing his father's claim to be 'father of the steam locomotive' to anyone he thought might listen. The Family Archive contains a collection of similar letters from John, written on behalf of his father to newspapers and technical journals. He was particularly affronted by the, unquestionably biased, biography of Stephenson by Samuel Smiles, who relegated Timothy to the role of bit player in railway history. In a typical undated diatribe to the *Northern Daily Express*, written sometime around 1858, John ranted that not only was Hackworth, and Hackworth alone, the inventor of the locomotive blast pipe, but that most of Samuel Smiles references to his father in Stephenson's biography were fictitious. He went on to do a similar hatchet job on the biography of William Hedley recently written by Hedley's son Oswald, which in John's view wrongly credited William Hedley, the Wylam viewer, and not Timothy Hackworth, for the pioneering locomotive work at Wylam. Hedley he ranted, 'was no mechanic, neither did he invent any part of the locomotives employed there…'

Timothy's later years were troubled by illness but his enthusiasm for steam engines never wavered. Not long before he died in 1851, in a letter to his son John that touchingly ends 'pray for me amen' he wrote;

> I believe myself that something more will be done, the Locomotive Engine is capable of vast improvement; … my own opinion is that something different from anything yet invented will be brought bear, in a much simpler manner than anything yet in use – simplicity & economy must be the order of the day.

In terms of his contribution to the development of the steam locomotive, he could have no better epitaph.

CHAPTER THIRTEEN

The Wylam Boom Years

Meanwhile, back in Wylam, plans were afoot to recycle the colliery spoil heaps – or at least their iron content. An iron works was built by Benjamin Thompson close to the village centre on land close to the ford over the Tyne. This opened for business in 1835 and was run by Thompson's two sons who owned collieries nearby. The proximity to the N & CR prompted the creation of an associated engineering works. It was here four 0-6-0 and two 2-4-0 steam locomotives were built for the N & CR between 1839 and 1841. It is more than likely that the 'Thompson' in the firm 'Hawks and Thompson' of Gateshead who were also building engines for the N & CR was part of the same family since the designs of both sets of locomotives, according to Tomlinson, had many features in common. No drawings of the Wylam engines exist but a drawing of the 0-4-0 Hawks and Thompson engine *Lightning*, built around the same time, gives some idea of what they might have looked like and may be compared with the engine shown in the engraving at the beginning of this book.

In 1844 the iron works was purchased by the Bell brothers who went on to found an iron works at Port Clarence on the River Tees. This was the time of greatest prosperity for Wylam. Its population, which was around 100 in 1700 reached its peak of a thousand plus. The centre of the village shifted east as clusters of new homes grew up around the iron works. A tramway/road bridge was built over the Tyne to link the works to the station on the N & CR; the same bridge that now conveys road traffic over the river. By now the stars of Wylam's prodigal sons were going supernova. George Stephenson's success with the Stockton and Darlington railway was followed by arguably his greatest triumph – the building of the Liverpool and Manchester Railway (L & MR) three years later. It had a shaky start. Employed as a railway engineering expert George was asked to give evidence to Parliament in support of the Railway Bill. Unlike the S & DR, where George knew every stone, nook and cranny along the proposed route, in this instance he left much of the survey work – and more importantly some of the crucial technical detail – to subordinates. When he then presented his evidence to the House, the Bill's opponents tore poor George to shreds. His bluff northern manner cut no ice with soft southerners. Even his attempts at humour

Hawks and Thompson's *Lightning*.

went down like the *Titanic*. At one point George was asked, given the weight and speed of locomotives, 'if a cow strayed on to the line would that not be a very awkward circumstance' to which George is said to have replied, 'Very awkward – for the coo.' Nobody laughed.

On question after question about how he intended overcoming the many formidable obstacles along the way George either prevaricated or provided vague or sometimes downright contradictory answers so, not surprisingly, the Bill was thrown out – much to the delight of a contingent of Lancashire canal owners who correctly enjoyed a cosy monopoly for transport of goods (particularly cotton) from Liverpool docks. [32] Despite Stephenson's mauling at Westminster the passing of the Bill was only delayed and, with the educated Sir John Rennie rolled out to talk 'proper' to Members of Parliament, the Bill finally gained Royal assent. It was, nevertheless, to the credit of the financiers of the L&MR that it was to George, rather than Rennie, they turned to as the engineer to build their railway. Rennie, to his obvious disgust, was relegated to the role of advisor, but soon withdrew after being told he was expected to work alongside the ignorant northerner. Stephenson still had a battle on his hands. Unlike the S & DR the terrain between Liverpool and Manchester was unrelentingly awful; immediately after leaving the docks, a long tunnel was needed followed by a vast deep cutting. Worst of all Stephenson had to negotiate the marshy morass of Chatmoss. There were many notable engineers of the day who said it couldn't be done and even George's normal optimism was tested to the limit. He had decided to float the railway over the quagmire by supporting the track on a raft of wooden hurdles with the hope that the underlying waterlogged peat would be displaced in the same way that a ship floats on water. The problem was that the more ballast he laid down, the more the track bed sank into the bog. Just when his employer's patience was running out, the structure finally stabilised and a flooded cutting became a dry embankment on which he could finally lay down rails.

With the L & MR nearing completion there remained the question of whether or not to use steam locomotives and, if so, which engines to use. That the proprietors chose to decide who got the contract for supplying engines only after a free-for-all trial, says much about the perceived reliability of Stephenson's S & DR engines at the time. Everyone now knows the outcome of the Rainhill trials with Stephenson's *Rocket* seemingly wiping the floor with the opposition, but at the time it was actually a close run thing. Nevertheless *Rocket* possessed a simple elegance that the others lacked. Its key features were a multi-tube

George and Robert Stephenson's home at Killingworth.

boiler and direct linkage between the cylinders and the driving wheels; with none of the levers, cogs and cumbersome linkage associated with earlier engines. Its main opponent *Sanspareil*, sadly, now looked a relic from the past. Despite its tiny dimensions, *Rocket* exhibited all the salient features of modern steam locomotives and it was no surprise when it was chosen to head the cavalcade of trains on the opening day of the railway that was paraded before by the Duke of Wellington. It did blot its copybook slightly by knocking down and killing the amiable MP for Liverpool and passionate railway advocate, William Huskisson; an incident which to be truthful put a damper on the day's proceedings. The opening of the L&MR, nevertheless, signalled the true arrival of railways and there was a sudden world-wide surge of interest.

The spotlight now shifted away from George Stephenson and on to his son Robert. New railways were springing up right across the UK and at the heart of the greatest and grandest was Stephenson junior. It was Robert who led the way in the subsequent development of locomotives and railways. He was fortunate to have learned the trade from the very best. The engineering and pioneering diligence of his father was now coupled to a sound education; received both as apprentice to the educated Nicholas Wood at Killingworth and more formally at a private school in Newcastle and as an undergraduate at Edinburgh University. Unlike George, Robert could talk the talk, taking care to moderate and eventually lose the Northumberland accent that gave his father so much grief at Westminster. He was even thought well of by Sir John Rennie, the very man who refused to work with George on the building of the L&MR. In the ensuing railway boom Robert became a key player designing and building some of the most magnificent engineering structures the world had ever seen, including a spectacular bridge over the Menai Straits between Anglesey and the Welsh mainland. His lasting legacy to his birthplace would be the design and construction of the High Level Bridge over the Tyne at Newcastle. With Robert, however, the Wylam connection was finally broken. Robert was born at Willington Quay and was now serving his aprenticeship alongside Nicholas Wood at Killingworth. The house where the Stephenson family lived in Killingworth is still standing; if a little incongruous in the middle of the bland modern housing estate that replaced the rows of terraced houses once occupied by mineworkers. A sun dial designed and built by father and son stands over the door.

At the time Robert's father was establishing his reputation as an inventor his friend and colleague Nicholas Wood had proved the perfect partner. Educated at all the right schools and well-spoken he was wheeled out by George whenever he needed a cultured voice to present his work to a sceptical establishment. Wood's name duly appeared on several of Stephenson's patents, including that of the 'Geordy Lamp'; George providing the invention and Wood the promotion. Although Wood would later make his name in the colliery industry for contributions to mining safety, contributions duly recognised by the Coal Mining institute who made him their first President, from a railway perspective he will always be remembered for his *Treatise*; a potted history of railways and a DIY guide on how to build them. This, originally published in 1825, but substantially revised in 1838, rejoiced in the wonderfully succinct moniker, 'A Practical treatise on Railroads and Interior Communication in General Containing Numerous Experiments on the Powers of the Improved locomotive Engines and the Tables of the Comparative Cost of Conveyance on Canals, Railways and Turnpike Roads'.

An unashamed promo for railways, it set out the logical and sound financial reasons why a railway was the best and cheapest (not to mention the coolest) method of transporting goods. It also provided a chronology of railways and a step by step guide to building them from scratch. In so doing it advanced the cause of railways almost as much as the experiments at Wylam and the contemporaneous efforts of George Stephenson and the Pease family.

Not surprisingly it is George Stephenson's overall contribution that figures most in Wood's *Treatise*. If Trevithick barely merits a mention William Hedley doesn't figure at all, except indirectly in references to the wheel/rail adhesion experiments and the building of an unidentified Wylam locomotive for which Hedley's boss, Christopher Blackett, is given sole credit. *Grasshopper*, if indeed that is the Wylam engine that gets a mention, is called 'Blackett's engine' and mentioned only because of its erratic behaviour; 'the irregular action of the single cylinder produced jerks in the machinery and had a tendency to shake the machine in pieces'.

There is no direct evidence Wood ever visited Wylam during the heyday of *Puffing Billy* or *Wylam Dilly* as he seems to have relied heavily on Stephenson's testimony of events.

According to his treatise, Wood conducted his own experiments at Killingworth to justify the claims he made in the book about the superiority of rail over other modes of transport. In 1818 Woods and Stephenson even repeated Hedley's adhesion experiments and reached the same conclusions. Wood's support for 'moving' over 'fixed' engines in the Treatise was timely given that both the S & DR and L&MR were on the verge of abandoning steam locomotives altogether, since on the available evidence they seemed less reliable and less efficient. The idea of linking the great cities of Liverpool and Manchester with winding engines and long ropes pulling trains, was even for a time the preferred option – as evidenced by the following letter from Robert Stephenson in the Hackworth Archive:

> The reports of the Engineers who visited the North to ascertain the relative merits of the two systems of Steam machinery now employed on Railways, have come to conclusions in favour of Stationary Engines. They have increased the performance of Fixed Engines beyond what practice will bear out, and I regret to say they have depreciated the Locomotive Engines below what experience has taught us. I will not say whether these results have arisen from prejudice or want of information or practice on the subject. This is not a point which I will presume to discuss. I write now to obtain answers to some questions, on which I think they have not given full

information. Some of their calculations are also at variance with experiments that have come under your daily observation. The reports of the Engineers who visited the North to ascertain the relative merits of the two systems of Steam machinery now employed on Railways, have come to conclusions in favour of Stationary Engines. They have increased the performance of Fixed Engines beyond what practice will bear out, and I regret to say they have depreciated the Locomotive Engines below what experience has taught us. I will not say whether these results have arisen from prejudice or want of information or practice on the subject. This is not a point which I will presume to discuss.

Reliability was the key word here and it was Hackworth rather than Stephenson who was going to turn things round with the building of the efficient *Royal George* and its successor engines. His son, John, born at Walbottle, was now called on to play his own part in the world-wide promotion of steam. In 1836, at the tender age of sixteen, he went out to Russia to provide the engineering support to a new Hackworth engine, the first to be supplied or even seen in that country. His passport was personally endorsed by the Tsar – although the 'Wesley' part of his name was changed to William to avoid upsetting the Russian monarch's orthodox sensibilities. As cold as it gets in north-east England John wasn't prepared for the type of weather common to Eastern Europe. Consequently,

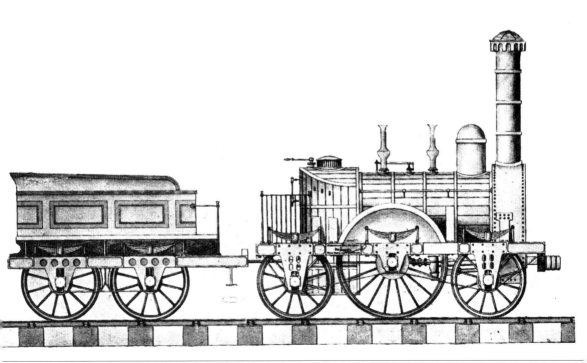

Hackworth's Russian locomotive.

the Russian port where he and his party had planned to land was found to be icebound and they were forced to dock at the first available ice-free Baltic port and make their way cross country, hauling the engine on a sledge. The problems associated with shifting an 8-ton locomotive across a snow bound landscape can only have been formidable; that it should be accomplished by someone so young says much for the tenacity, self-belief and faith of the Hackworth family. On its way to St Petersburg, John's party was attacked by a pack of wolves, not the sort of experience he would have been prepared for from an apprenticeship in County Durham. After making some repairs to the engine, which suffered on the journey due to the sub-zero temperatures, John paraded the engine before Tsar Nicholas at the latter's summer palace in Tsarkoye-Selo. It transpired the Tsar knew a fair amount about steam engines having seen Blenkisopp's rack-and-pinion locomotive at work in Leeds twenty years earlier. He was therefore able to give it his Royal seal of approval, complimenting John on the advancements made in locomotive technology since his last visit to England. There was a formal blessing of the engine before it was put into steam. The locomotive was surrounded by a ring of lighted candles and baptised by a priest before an assembled crowd. Each of the engine's wheels were individually anointed with holy water. Tsar Nicholas must have been impressed as he instituted a major programme of railway building to link all the major cities of Russia, funded by selling Alaska to the United States.

Many years later John Wesley Hackworth took over his father's business but specialised in stationary engines, particularly winding engines for the collieries and high efficiency 'horizontal' high pressure engines for the cotton industry, which were manufactured in his new works built near Darlington Bank Top station – where Locomotion No. 1 had at last come to rest.

Railways had taken off in a big way with everybody trying to get in on the act and shares trading at exorbitant prices for the most unlikely railway ventures. By the middle of the century the industrial revolution was approaching a summit while the fortunes of Wylam were declining. The first industry in the village to go was the iron works, which closed in 1864 when locally sourced iron ran out. The coal beneath the village was exhausted shortly after. Bit by bit the smoking chimneys of Wylam disappeared; the lead works, the brewery and then the people.

Epilogue:
The Wylam Legacy

It is now hard to believe that Wylam was once a thriving industrial complex. The entire mining infrastructure has gone – the workshops, the winding engines and the railroad. The spoil heaps that dominated the landscape have been grassed and graded and are now indistinguishable from natural undulations in the surrounding fields. The famous wagon-way connecting the colliery to the staiths at Lemington, the one that ran past George Stephenson's cottage, is now a Sustrans cycle path much frequented by joggers and dog walkers. Where the Tyne Valley echoed to the sound of industry there is now only birdsong, and the occasional whine of a jet heading for Newcastle Airport. For a time, as we know, the wagon-way became part of the national rail network, having been extended beyond the village and over the river on Hagg Bridge before re-joining the main line to Carlisle.

Until the 1960s it was possible therefore to catch a train from Newcastle to Carlisle, which went right past Stephenson's birthplace. The nearest station, North Wylam, a mile or so to the west, closed along with the rest of the northern loop and is now the village car park. South Wylam (now just plain Wylam), is a short saunter to the south on the far side of the bridge over the river, built to accommodate tramway trucks from the former iron works. The surviving Wylam station is, curiously, the more venerable of the two, being built for the Newcastle and Carlisle Railway and opening in 1835. It is therefore one of the oldest surviving original railway stations anywhere in the world.

If you walk along the river side, as I did in the company of local historian Denis Peel, you can still see coal seams outcropping on the river bank, but the coal mines that harvested the black gold are long gone, along with the iron works whose furnaces once turned the local sky blood red. Traces of industry are now hard to find; here the odd incongruous mound, there the odd wall made from furnace slag. The valley north of the village, where the stream was diverted so coal workings wouldn't flood, has disappeared; the coal dried up and the valley was used as a tip for disposal of iron works slag. Apart from traffic grappling with the narrow streets, Wylam today has more in common with its pre-industrial self than the industrial powerhouse it became in the early part of the nineteenth century. In

The Wylam wagon-way today (Stephenson's cottage is on the right).

Hagg Bridge at Wylam.

Epilogue: The Wylam Legacy

The Wylam end of the wagon-way (2011), the site of Haugh Pit.

two hundred years the wheel has turned full circle; Wylam is now what it was in medieval times – a rural village on a crossing point of the Tyne. During the hundred years the colliery thrived the population increased six fold reaching 673 in 1801 and over a 1,000 just before the iron ran out. Ten years later a quarter of the residents had moved away and another hundred years would pass before the population reached the numbers it achieved then.

There are occasional points of reference. Commemorative plaques have been placed by the local history society on buildings that occupy the sites of the former homes of Hedley and Hackworth, although the original cottages have been demolished. The Black Bull pub, built to serve the passing trade of drovers using the Tyne ford, and to later slake many a miner's thirst, is still serving ale and Hackworth's blacksmith's shop, where the metal parts for *Wylam Dilly* and *Puffing Billy* were forged, has become a craft shop, although virtually invisible in a housing estate built more than a hundred years later.

Between 1820 and 1860 the structure of the village changed completely. Rows of terraced houses were built close to the centre of the village to house employees of the iron works and the centre of the village moved to accommodate the change. There is no trace of the iron works that wrought the changes. A library now stands on the works site. It has a one room museum dedicated to Wylam's illustrious railway history. There I met Phillip Brooks, the Chairman of the

Hackworth's blacksmith's shop today.

History Society and the only biographer, other than the Wylam viewer's son and agent to champion the cause of William Hedley. The Haugh Pit where Hedley was based has become a grassy mound among trees. Nothing remains of the lead shot manufactory or sadly William Brown's brewery…alas…alas.

Street House, George Stephenson's birthplace is of course still there, standing next to the former wagon-way. It is now a Mecca for railway pilgrims from all over the world. If you should ever pay it a visit you will find a local volunteer, dressed in period costume, who explains what life was like for the family of seven that lived in that small downstairs room at the end of the eighteenth century. In Stephenson's day there were nineteen other people crammed into this tiny cottage. Today, with its attractive four poster bed and cooking pot hanging from a swinging spit over the fire, it all looks rather cosy – even desirable. The flag stone floor you stand on is a later edition. It was bare mud in George's day. As a tied cottage the only claims on occupancy the residents had was through employment by their landlords, the Blackett family, so when George's father Robert was laid off the family were evicted. It is possible to speculate therefore that if things had worked out differently it might have been Stephenson rather than Hedley who pioneered locomotive engineering at Wylam.

The last of Christopher Blackett's descendants, resident in Wylam, died in the 1970s and in a garden shed of his home (no longer Wylam Hall) a jumbled

Epilogue: The Wylam Legacy

Street House at Wylam today.

heap of boxes of family correspondence was discovered, dating back 200 years. This was parcelled up and shipped to the Northumberland County Record Office at Woodhorn where it resides today. It has yet to be fully catalogued but there appears to be barely a shoe box full of papers relating to the crucial years 1800 to 1820 when the historic locomotive construction was taking place. Even such correspondence as exists, dating from that time, relates mainly to family matters, particularly the moving letters from Alice, Christopher's wife, to her son Christopher expressing her concerns about his safety in Wellington's army. There is no mention of her husband's work.

After William Hedley moved to Burnhopeside Hall, his involvement in the family's many commercial interests dwindled. Two of his sons took over the colliery business, operating under the name Thomas Hedley & Brothers, with a head office on the quayside in Newcastle. Of the three locomotives with which Hedley is directly associated two outlived him. *Puffing Billy* worked on at Wylam until the mid-nineteenth century, being rebuilt so many times it has since proved difficult identifying the original features. It was eventually sold by Christopher Blackett's son in 1862 to the London Patent Office [33] and is still an incongruous attraction in the Science Museum, located among modern relics of the space age. *Wylam Dilly*, though modified frequently, lasted even longer. It continued in use at Wylam until the colliery closed in 1868 [34] whereupon it was

Puffing Billy in its final four-wheeled state.

Epilogue: The Wylam Legacy

Puffing Billy at the Science Museum.

George and William Hedley (in top hats) at Craghead Colliery.

purchased by Hedley's sons for the scrap value of £16 10s and moved to their Craghead Colliery. A much reproduced photograph exists showing Hedley's sons George and William stood beside *Wylam Dilly* at Craghead. They are decked out in top hat and tails; a striking contrast to the driver and fireman dressed in grimy labourers togs nearby. Later portraits of Hedley himself show him similarly arrayed and clearly the family had moved up in the world and become 'landed gentry'. Burnhopeside Hall is the surviving manifestation of the family's change in circumstance. The boy Hedley had 'done well' for a grocer's son [35]. As an eminent pillar of the local establishment Hedley was invited to become a member of the prestigious 'Literary and Philosophical Society' at Newcastle where he regularly gave lectures on locomotive engineering and railways in general, at a time when, according to his agent, 'their [railways] modern meaning was all but unknown'. Since Wylam never possessed a parish church, Hedley's son George paid for the future St Oswins to be built in the village and dedicated it to the memory of his father. Another of Hedley's sons, William, paid for the nearby vicarage.

In the years that followed there would be many claimants for the title 'Builder of the first successful steam locomotive', however, one man, George Stephenson, has been lauded above all the others, not least by his biographer Samuel Smiles, who dismissed the claims of all Stephenson's predecessors. William Hedley had to rely on the spirited defence of his son Oswald and his agent Mark Archer to promote his own claim. Neither biographers can be considered objective, even if Archer devoted half his book to providing independent advocates of Hedley's case; summarising his conclusions thus, '[Hedley] constructed the first practically successful and economically working locomotive engine', and with an inevitable sideswipe at the Stephenson camp, 'In 1828 Stephenson's engines were looked on with disfavour while HEDLEY's [his capitals] had all along been working satisfactorily'.

This was, despite Archer's bias, substantially correct. Until Hackworth started building locomotives for the S & DR that railway, as we know, was seriously considering returning to horses for motive power purely because of the unreliability of Stephenson's engines. Hackworth's descendants likewise dismissed Hedley's contribution to the success of the Wylam engines. As late as 1920 Hackworth's direct descendant Samuel Holmes told Robert Young, Hackworth's biographer and grandson, to ignore Hedley completely, 'Blackett, Blackett, Blackett not Hedley', he insisted.

Robert Young's father, George, compiled his own hand written account of early railways with the emphasis on Timothy Hackworth as the true 'father of railways'. The reference to Wylam in the book for example contains the unsubstantiated claim that the Wylam engines were, 'built in 1812 or 1813 by Timothy Hackworth and Jonathon Forster'.

The most aggressive promotion of Hackworth's place in locomotive history, however, comes from a series of letters to journals by Timothy's son, John Wesley Hackworth, who not only attacked George Stephenson, who he claimed copied the primitive blast pipe on *Puffing Billy* for his own engines and didn't build a reliable locomotive until Hackworth moved to the S & DR, but also went out of his way to do a similar hatchet job on Hedley. It is worth reproducing his diatribe in full:

> And now Mr.Editor I certainly give you credit for honesty of intention in arguing the claims of William Hedley but I presume you had no personal acquaintance with him whereas I did know him also his predilection – as viewer – to the introduction of the locomotive at Wylam. However he was no mechanic, neither did he invent any part of the locomotives employed there as I proved in a discussion many years ago. Both at the time his son Oswald Dodd Hedley issued his book entitled 'Who Invented the Locomotive Engine' in 1858 and afterwards when the Times 9in January 1865) had an article on Patriarchal Engines at South Kensington and although the Times excluded my answer yet as the local, papers copied the article I obtained an opportunity to correct certain misstatements therein....

Since John wasn't born at the time Hackworth was working at Wylam, his claim to know Hedley should be treated with more than a little caution. Unfortunately, as we all know, history is written by the survivors and blacksmith Hackworth, unlike Hedley, was followed by an army of descendants with axes they were desperate to grind.

So what exactly was the input of each of the Wylam pioneers? Well, of the four key players in locomotive construction: Blackett had the vision and the money;

Hackworth was the skilled blacksmith who forged the components; Forster (not Hackworth it should be noted) was the mechanic who assembled them and made them work; and finally, Hedley designed the engines and made sure everything happened in time and on budget the way Blackett wanted it to.

Both Philip Brooks and Denis Peel believe that Hedley's personal significance has been overlooked by history. The recent deification of Timothy Hackworth, because of his important development work in early locomotive engineering, has led to the widespread view that Hedley was just an onlooker to the momentous events at Wylam. Hackworth and Jonathon Forster, it is argued, were the main players. Hedley does not even merit a mention in Nicholas Wood's famous treatise on railways. The view that Hedley was a peripheral figure in the creation of the Wylam engines seems to me to be a distortion of reality. A colliery viewer's overarching responsibility was to serve the interests of the colliery's employer. Blackett, who does receive credit in Wood's treatise, was far sighted enough to see the future of steam locomotion but it was up to Hedley to make the dream reality. Of the triumvirate who worked on the Wylam engines Hedley should be considered as having an analogous place to Wood's role at Killingworth when Wood assisted the less well educated Stephenson with locomotive design and construction; at Wylam Hedley provided the technical input and literary skills that complimented the practical abilities of Forster and Hackworth.

A glimpse of Wylam *c.* 1813 (*Puffing Billy* at Beamish).

Epilogue: The Wylam Legacy

In October 1882 the Hedley family exhibited Wylam Dilly at the 'North East Coast Exhibition of Naval Architecture and Marine Engineering' at Tynemouth and while it was on display there it was donated by Hedley's agent Mark Archer to the Royal Scottish Museum for permanent display in Edinburgh, where it remains to this day. *Puffing Billy* was purchased initially for the Patent Office then moved to the Kensington – now Science – Museum along with Hedley's model of the wheel adhesion test carriage. The ultimate fate of *Lady Mary* has never been adequately resolved, which as we have seen, has led to speculation that it never existed at all.

Unlike George Stephenson, who is rightly feted throughout the world, the only acknowledgement of Hedley's contribution to steam locomotive development was the naming of a 2-4-0 tank engine *Hedley* by the Great Western Railway in 1865. Ironically Hedley's iron namesake was scrapped in 1929, on the 150th anniversary of his birth. Hedley was buried in the family plot at Newburn Parish Church, in the same church where he was married. The 'White House', in which he lived during his time at Wylam, was demolished at the turn of the twentieth century.

The full sized working model of *Puffing Billy* lives on at Beamish where its working replica provides entertainment for throngs of day trippers during the summer months. The Beamish 'Billy' was built by Alan Keefe at Ross-on-Wye for £500,000, over a thousand times more than its illustrious forebear. It is equipped with many safety features *Puffing Billy* never possessed including air brakes and a steam pressure gauge. The boiler, like that of Stephenson's *Rocket*, is multi-tube and the metal structure reinforced. Looking at the reconstructed locomotive pulling rowdy parties of schoolchildren back and forth along the reconstructed wagon-way, it is hard to picture its predecessor all those years ago – when Wylam village was the railway centre of the world.

At the beginning of this book I wondered what combination of circumstances and events turned pitmen into pioneers at Wylam. I am unsure whether there is a definitive answer. The need to provide a cheap alternative to the horse for transporting coal and just being in the right place at the right time is part of the answer, but why it happened at Wylam, in preference to anywhere else, can best be countered by the words 'why not?' The world was changing and what happened in this small Northumberland village would probably have happened somewhere else eventually. The steam locomotive was by then a reality and its potential for haulage was well established, even in 1810. However, Wylam made the use of steam locomotives for transportation seem 'normal'. It was this fact rather than any specific invention that led to the exponential development of railways. Wylam showed that railways could work and for that we must all be grateful.

Acknowledgements

Alison Kay and the National Railway Museum, for permission to reproduce extracts from the Hackworth family Archive; Dr Jill Murdoch and the Institute of Railway Studies, University of York; The Literary and Philosophical Society at Newcastle; The National Archive at Kew; Beamish Industrial Museum, County Durham; The Northumberland County Council historical archive at Woodhorn; Phillip Brooks and Denis Peel of Wylam History Society; and The Mining Institute in Newcastle.

All the modern images are courtesy of the author. The author acknowledges the debt owed to the railway historians of the past. The locomotive diagrams and early railway drawings are mostly taken from original engravings as reproduced in Nicholas Wood's 1825 *Treatise of Railways*, Robert Young's 1925 biography of his grandfather, Timothy Hackworth, Clement Stretton's 1896 book *The Development of the Locomotive* and Tomlinson's 1925 book about the North Eastern Railway. The portraits of the elder Hedley and Hackworth are also reproduced from Young's book. The 'Hedley Test Carriage' is a Science Museum image, subsequently reproduced in numerous railway books including Brooks' short biography of Hedley. The 1808 portrait of Hedley is taken from a portrait reproduced in Brooks' biography as is the 'keel boat *Dilly*', which was originally a contemporary watercolour. The picture of North Wylam station in the 1950s is courtesy of Alan Young.

References

1 Built by William and Alfred Kitching at the Hope Town Foundry in Darlington in 1845.

2 The wage for a coal 'hewer' at Tanfield in 1728 was between 1 shilling and sixpence and 1 shilling and ten pence per day (between 5 and 8 pence).

3 Not be confused with the Geordie expression 'hooers', which is a different kettle of fish altogether.

4 Also spelt 'Murdock'.

5 It wouldn't be long before these unwanted impurities came into their own when it was discovered that coal gas made an acceptable fuel in its own right and everything from coal tar to ammonia based fertilisers would soon be made from the other 'unwanted' volatiles.

6 John Steel may have worked for Whinfield at Gateshead before he went to Pen-y-Darren. Steel would meet an unfortunate end, being blown up by one of his own engines while working in France.

7 They were then known as 'cinder ovens'.

8 'Dodds' is a name closely associated to with the coal mining industry. Ralph Dodds, for example, was the viewer at nearby Long Benton Colliery.

9 It was a good time to own a collier boat in the north east. Because of the on-going war, most of the large colliers had been commissioned by the Army for freight transport to the Baltic.

10 Although in one of his letters written in 1849, now in the Hackworth Family Archive at the NRM, Hackworth refers specifically to working on

'locomotive construction for 39 years' making the year 1810 – not 1911 (ref. HACK 1/1/35).

11 It is a good thing the Iron Duke didn't charge royalties for the toadying use of his name.

12 The idea of a chain operated railway between Liverpool and Manchester was seriously considered by that particular company's directors before they actually committed themselves to steam locomotives.

13 In fact, the patent was subsequently challenged for having no novel elements and was later rescinded.

14 Hedley's agent, Mark Archer, also referred to it as *Puffing Dilly*.

15 A fact that Hedley's son Oswald emphasised when pursuing his father's claim as being the inventor of the blast pipe.

16 Losh would be a future director of the Newcastle and Carlisle Railway.

17 In fact, Lowe suggests that Chapman actually built an eight-wheeler for Wylam Colliery, although there is no other evidence for this.

18 In 1827.

19 The same pit which was used for the disposal of bodies during the Jacobite Wars.

20 In fact, the original keel boat that was adapted to take the Dilly was called for some unexplained reason *Tom and Jerry* – 'Looney Tunes' indeed.

21 Since he was also called Christopher and would eventually become the Wylam Colliery owner, he is often confused with his father in railway history books e.g. *The Great British Railway Living History* by Tony Hall-Patch.

22 He had even produced a booklet promoting the use of his locomotives in other collieries.

23 The 'Surrey Iron Railway' from Wandsworth to Croydon.

24 Although Smiles calls it the *Active*.

25 The design was not coincidental. The engineer who built the *Agenoria* was John Rastrick, a native of Morpeth in Northumberland, who would have seen the Wylam engines at work.

26 Or at least a son of nearby Prudhoe!

27 This was originally intended for use on the Canterbury & Whitstable Railway (C & WR), but was subsequently surplus to requirements when the C & WR decided only to use steam locomotives on their rails.

28 Named after the current monarch George IV and not the then out of favour George Stephenson.

29 Although Trevithick had designed one in 1815 according to Samuel Smiles – that may have been the model used for *Rocket*.

30 Hackworth's descendants even rescued one of Stephenson's former apprentices from the workhouse so he could provide written testimony to the effect that the piston was deliberately sabotaged before it left Forth Street.

31 History seems to have recorded little about Mr Downing. He seems to have been a sleeping partner in the firm, probably putting up the finance for the operation.

32 Part of the problem was the intransigence of the landowners over whose land the railway was to be built. Surveying work had to be conducted much of the time in darkness with the ever present risk of getting shot.

33 In fact he only loaned it in the first instance as he had just acquired a new locomotive to replace it, suggesting that *Puffing Billy* was still doing daily work at the colliery until then.

34 Although Dendy Marshall suggests that two pits at Wylam continued operating until 1880 'when they became flooded, when another pit opened up nearby that operated until 1923'.

35 A John Hedley, presumably Hedley's brother, is listed in the *Northumberland & Durham Gazetteer* as still being the only grocer based in Newburn village in 1828.

Principal Bibliography

Archer, M., *William Hedley, The Inventor of Railway Locomotion on the Present Principle* (J. M. Carr, Newcastle 1882).

Extracts from the *Blackett Family Archive* at Woodhorn.

Bleasdale, P. E. M., *Railway Magazine*, 'William Hedley and the Wylam Locomotives', (June 1927), pp. 445–452.

Brooks, P. R. B., *William Hedley Locomotive Pioneer*, (Tyne and Wear Industrial Monuments Trust, Newcastle, 1980).

Burton, A., *Richard Trevithick – Giant of Steam*, (Aurum Press Ltd, 2000).

Crompton, J., 'The Hedley Mysteries', *Paper presented to the Second International Early Railways Conference* (2003); reproduced in *Early Railways*, 'Vol. 2'.

Davies, H., *George Stephenson – A Biographical Study of the Father of the Railways*, (Quartet Books Ltd, 1977).

Dendy-Marshall, C. F., *History of the Railway Locomotive down to the year 1831*, (The Locomotive Publishing Company Ltd, 1953).

Dodds, M. H., *A History of Northumberland*, 'Vol. 12', (1930).

Engels, F., *The Condition of the Working Class in England*, (Germany, 1845) – edition used for reference Penguin Classics, 1987.

Flinn, M. W., 'The Industrial Revolution', in *The History of the British Coal Industry*, 'Vol. 2', 1730–1830, (Clarendon Press, Oxford, 1984).

Flinn, M. W., *Origins of the Industrial Revolution*, (Longman Group Ltd, 1966).

Fynes, R., *The Miners of Northumberland and Durham*, (Thomas Summerbell, 1873).

Extracts from *The Hackworth Family Archive* at *The National Railway Museum*, York.

Hedley, O. D., *Who Invented the Locomotive Engine*, with a review of Smiles' 'Life of Stephenson'.

Lewin, H. G., *Early British Railways – A Short History of Their Origin and Development 1801–1844*, (London, *Railway Gazette*, 1936).

Lowe, J. W., *British Steam Locomotive Builders*, (Goose and Sons, 1975).

Ottley, G., *A Bibliography of British Railway History*, (George Allen & Unwin Ltd, 1965).

Reynolds, M., *The Model Locomotive Engineer, Fireman and Engine Boy*, (Crosby, Lockwood & Co., 1879).

Skeat, W. O., *George Stephenson – the Engineer and His Letters*, (Institute of Mechanical Engineers, 1973).

Smiles, S., *The Life of George Stephenson, Railway Engineer*, (John Murray, London, 1857).

Stratton, C. E., *The Development of the Locomotive (A Popular History 1803 – 1896)*, (Crosby Lockwood and Sons, London, 1896).

Stretton, C., *The Development of the Locomotive*, (London, 1896).

Tomlinson, W. W., *Tomlinson's North Eastern Railway* (first published in 1914; reproduced with introduction by K. Hoole in 1967. Third Edition (used for reference here) published by David & Charles Publishers PLC of Devon, 1987).

Welford, R., *Men of Mark Twixt Tyne and Tweed*, 'Vols. 1 – 3', (Walter Scott Ltd, Newcastle, 1895).

Wood, N., *A Practical Treatise on Railroads and Interior Communication in General – Containing Numerous Experiments on the Powers of the Improved*

Locomotive Engines and Tables of the Comparative Cost of Conveyance on Canals Railways and Turnpike Roads (sic), (Longmans, Orme, Brown, Green and Longmans, 1838).

Young, R., *Timothy Hackworth and the Locomotive*, (Locomotive Publishing Co. Ltd, 1925).